The Atomic Bomb
in Images and Documents

The Atomic Bomb in Images and Documents

The Manhattan Project and the Bombing of Hiroshima and Nagasaki

SAMUEL S. KLODA

McFarland & Company, Inc., Publishers
Jefferson, North Carolina

LIBRARY OF CONGRESS CATALOGUING-IN-PUBLICATION DATA

Names: Kloda, Samuel S., 1948– author.
Title: The atomic bomb in images and documents : the Manhattan Project and the bombing of Hiroshima and Nagasaki / Samuel S. Kloda.
Other titles: Manhattan Project and the bombing of Hiroshima and Nagasaki
Description: Jefferson, North Carolina : McFarland & Company, Inc., Publishers, 2022 | Includes bibliographical references and index.
Identifiers: LCCN 2021057651 | ISBN 9781476684888 (paperback : acid free paper) ∞
ISBN 9781476643311 (ebook)
Subjects: LCSH: Manhattan Project (U.S.)—History. | Atomic bomb—History—20th century. | BISAC: HISTORY / Military / Nuclear Warfare
Classification: LCC QC773.A1 K56 2022 | DDC 355.8/25119097309044—dc23/eng/20220112
LC record available at https://lccn.loc.gov/2021057651

BRITISH LIBRARY CATALOGUING DATA ARE AVAILABLE

ISBN (print) 978-1-4766-8488-8
ISBN (ebook) 978-1-4766-4331-1

© 2022 Samuel S. Kloda. All rights reserved

No part of this book may be reproduced or transmitted in any form or by any means, electronic or mechanical, including photocopying or recording, or by any information storage and retrieval system, without permission in writing from the publisher.

Front cover images: *clockwise from top left* Aerial view of mushroom cloud, which rose to 50,000 feet (Wikimedia Commons). Letter from Stimson to Truman, declassified April 12, 1974 (Harry S. Truman Library). Photograph of ground zero before the atomic bombing of Nagasaki, Japan; the stadium was the aiming point; autographed by Abe Spitzer, radio operator (U.S. Air Force Photograph/Collection of Samuel S. Kloda). *Enola Gay* crew autographed by Tom Ferebee, Bombardier; Paul Tibbets, pilot; Theodore Van Kirk, Navigator (U.S. Army Photograph/collection of Samuel S. Kloda)

Printed in the United States of America

McFarland & Company, Inc., Publishers
Box 611, Jefferson, North Carolina 28640
www.mcfarlandpub.com

Table of Contents

Acknowledgments — vii
Preface — 1

1. Uranium and Plutonium — 5
2. The Manhattan Project — 8
3. Los Alamos National Laboratory — 21
4. The First Atomic Bomb: Trinity — 27
5. Paul W. Tibbets — 46
6. Preparations for the Atomic Bombings — 50
7. Tinian — 62
8. The Potsdam Declaration — 68
9. Planning the Atomic Bombing of Hiroshima — 77
10. The Atomic Bombing of Hiroshima — 94
11. Truman's Address to the Nation — 123
12. The Atomic Bombing of Nagasaki — 126
13. Japan Surrenders — 154
14. The Bikini Atoll Tests — 162

Epilogue: Japan — 179
Bibliography — 209
Index — 211

Acknowledgments

I want to express my sincere gratitude to the people who helped me with this book.

First, to my niece, Dr. Lorie Kloda, associate university librarian at Concordia University, for her valuable comments and for editing the final draft.

To Geneviève Beaudry, archivist at McGill University Library, and to Yves Lapointe, director and university archivist at McGill University.

To Laurie Austin, Audiovisual Archives of the Harry S. Truman Library, for her dedication to my project.

I am indebted to several institutions for their help in obtaining items for my collection and book.

To Mr. Yuichi Yokoyama, Curatorial Division of the Hiroshima Peace Memorial Museum, for his research into the LeMay leaflets. These leaflets are extremely rare, and their presence in my book is greatly appreciated.

To the Air Force Historical Research Agency, for its help in finding missing documents.

To Elizabeth Otto of RR Auctions for providing several treasures for my book.

To those at the Library of Congress for their tireless work.

To Daniel Meyer, director, Special Collections Research Center, University of Chicago, for finding beautiful Los Alamos photos for me.

To Lynda Corey Claassen, director, Special Collections & Archives, UC San Diego Library, for her help in obtaining the Einstein/Szilard photo.

To Adam Berenbak, archivist, Center for Legislative Affairs, National Archives and Records Administration.

To Rebecca Percz, administrator, Strategic Communications, Lawrence Berkeley National Laboratory.

To Erin Clements Rushing, outreach librarian, Smithsonian Libraries.

To Jennifer Loredo, U.S. Army Heritage and Education Center.

To Eric Slander, Textual Reference Archives II Branch (RR2RR), National Archives at College Park, Maryland.

Thank you to two wonderful newspapers: the *Lansing State Journal*, for allowing me to use the interview of Richard Thelen, and the *Santa Fe New Mexican*, for letting me reproduce its August 6, 1945, issue.

Finally, my thanks go to the members of the 509th Composite Group, who were kind enough to sign all my photographs and recount their stories over a period of 40 years.

Preface

In 1975, I graduated from Concordia University in Montreal, Canada, with a master's degree in science, with a specialty in partial differential equations. Three years later, in 1978, I started to do some part-time work as a mathematician in Washington, D.C. The best part of my job was being invited to interesting parties and going to lectures of famous people.

On one such occasion, I listened to a talk by General Paul Tibbets, the pilot of the *Enola Gay*. I had always collected autographs, but I had nothing for him to autograph that day, and worse, I didn't have my camera with me. However, General Tibbets gave me his business card, and the rest, as they say, is history. I went to the United States Air Force archives and found several photos of Colonel Tibbets with the *Enola Gay*. These I promptly got autographed.

General Tibbets always answered my questions about all facets of the two atomic missions and his involvement in the Bikini Atoll tests. He also helped me obtain the addresses of the other crew members and invited me to the 509th Composite Group's biennial reunions.

My correspondence with General Tibbets lasted for almost 30 years. I am very grateful for his help and for directing me to many of the crew members. General Tibbets played a vital role from the very beginning of the Manhattan Project to the Bikini Atoll tests.

Two of my friends in Washington, D.C., were Rita and Ben Gilman. Rita was a lawyer and knew everyone in Washington. Ben was a Republican member of the United States House of Representatives from 1973 until 2003. During his time in Congress, he was chair of the House Committee on International Relations (104th through 106th Congresses). He served as a sergeant with the Army Air Corps during World War II, flying 35 bombing missions over Japan.

Representative Gilman introduced me to William Colby (former head of the CIA). After telling Mr. Colby about my atomic bomb autograph collection, he directed me to the archives, where I obtained better photographs and documents dealing with the atomic bombing of Hiroshima and Nagasaki.

So, after collecting for 40 years, I have amassed approximately 900 photographs and declassified documents dealing with the atomic bombing of Hiroshima and Nagasaki, all autographed by the scientists who built the atomic bombs and the flight crews that delivered the bombs. In February 2020, I agreed in writing to donate my complete collection to the University of Chicago and the University of New Mexico.

Over the years I gave numerous talks on my collection and the story of the atomic bombings. At the end of my talks, after fielding questions, people would always come up and ask to see some of my photographs. I was reluctant to hand over these items, because they were all autographed and I was afraid they would be damaged in some way. In time, however, I realized that nearly everyone invited to these lectures had never seen an autographed picture or a declassified document.

Another common thread in most of the books I read about the atomic bomb was a lack of rigorous research. It seems that authors relied on previous writers for their information without actually checking the source. A case in point is the date when President Harry Truman actually found out about the Manhattan Project. Almost every book written on this topic states that as soon as President Roosevelt died on April 12, 1945, Truman was told about the Manhattan Project. This is not true.

My archival search in the Harry S. Truman Library revealed a letter from Secretary of War Henry L. Stimson to President Truman. The letter is dated April 24, 1945. Stimson asked the president for a meeting to "talk with you as soon as possible on a highly secret matter." The next day, in the presence of General Leslie Groves, Stimson informed the president about the Manhattan Project and the atomic bomb.

It is for these reasons that I decided to write my book containing archival photographs and documents dealing with the atomic bombing of Hiroshima and Nagasaki.

Fortunately, I had clearance to visit numerous archives and requested items that I wanted for my collection. Whenever a document was listed as "classified," I would ask for declassification under the Freedom of Information Act or simply write to the organization responsible for a classifying review.

I strongly believe when a person reads a declassified document, that person will feel some of the tension of the original writer and see that every word of the document was important.

In trying to explain the theory behind an atomic bomb, it was necessary for me to start right at the beginning of the discovery of uranium and plutonium. Just as the study of advanced calculus starts with a basic understanding of arithmetic, the building blocks of the atomic bomb start with the theory of atoms.

It is really quite remarkable how the study of atoms has progressed so fast. On December 11, 1922, when Professor Niels Bohr received the 1922 Nobel Prize in Physics, he made the following opening statement: "The present state of atomic theory is characterised by the fact that we not only believe the existence of atoms to be proved beyond a doubt, but also we even believe that we have an intimate knowledge of the constituents of the individual atoms." Less than 23 years later, the world saw the dawn of the atomic age.

In this book, I trace the six-year story from the formation of the Manhattan Project through the atomic bombings of Hiroshima and Nagasaki, concluding with some of the important atomic tests that followed.

Preface

In chapters 1 through 4, I explain the origins of Manhattan Project and the first atomic test at Alamogordo, New Mexico. Chapters 5 through 10 deal with the preparations and dropping of the atomic bombs. In chapter 13, I discuss the details of the Japanese surrender. I have also included details of the Bikini Atoll tests and an epilogue.

Throughout this book, I have included photographs and documents (in most cases autographed by the individuals involved) that complement the descriptions in the text.

Chapter 1

Uranium and Plutonium

Our story begins with a brilliant German chemist named Martin Heinrich Klaproth. Besides discovering uranium in 1789, he went on to discover zirconium, titanium, strontium, cerium and chromium.

It was not until 1898 that the new element of uranium was studied more carefully. In that year, a New Zealander named Ernest Rutherford was appointed to the chair of physics at McGill University in Montreal, Canada.

In 1903, Rutherford published "Radioactive Change," in which he stated that atoms could be transformed and that each atom potentially carried a tremendous amount of energy. Rutherford also demonstrated that radioactivity was the spontaneous disintegration of atoms. For this monumental discovery, he was awarded the 1908 Nobel Prize in Chemistry.

In 1907, Rutherford went to the University of Manchester, where he directed the Geiger-Marsden experiment that discovered the nuclear nature of atoms. In 1911, Rutherford began firing alpha particles at a thin sheet of gold. The alpha particles should have passed straight through, but some bounced back. This result meant that the atom was not solid. It was mostly empty space with a mass at the center—a tiny nucleus.

After these successful particle-scattering experiments, Rutherford himself was astonished at the results. In his book *Rutherford and the Nature of the Atom*, Edward da Costa Andrade quotes Rutherford as saying, "It was quite the most

Ernst Rutherford, New Zealand–born British physicist who came to be known as the "Father of the Nuclear Age." Through experimentations that started at McGill University, Rutherford discovered that nearly all of the total mass of an atom is concentrated in a nucleus. Eventually, this discovery led to the invention of the atomic bomb. In 1908, Rutherford won the Nobel Prize in Chemistry. Otto Hahn, who would later discover atomic fission, worked under Rutherford at the Macdonald Laboratory in Montreal (W.H. Hayles/McGill University Archives, PR017771).

(Reprinted from NATURE, July 7, 1923.)

The Structure of the Atom.[1]
By Prof. N. Bohr.

The General Picture of the Atom.

THE present state of atomic theory is characterised by the fact that we not only believe the existence of atoms to be proved beyond a doubt, but also we even believe that we have an intimate knowledge of the constituents of the individual atoms. I cannot on this occasion give a survey of the scientific developments that have led to this result; I will only recall the discovery of the electron towards the close of the last century, which furnished the direct verification and led to a conclusive formulation of the conception of the atomic nature of electricity which had evolved since the discovery by Faraday of the fundamental laws of electrolysis and Berzelius's electrochemical theory, and had its greatest triumph in the electrolytic dissociation theory of Arrhenius. This discovery of the electron and elucidation of its properties was the result of the work of a large number of investigators, among whom Lenard and J. J. Thomson may be particularly mentioned. The latter especially has made very important contributions to our subject by his ingenious attempts to develop ideas about atomic constitution on the basis of the electron theory. The present state of our knowledge of the elements of atomic structure was reached, however, by the discovery of the atomic nucleus, which we owe to Rutherford, whose work on the radioactive substances discovered towards the close of the last century has much enriched physical and chemical science.

According to our present conceptions, an atom of an element is built up of a nucleus that has a positive electrical charge and is the seat of by far the greatest part of the atomic mass, together with a number of electrons, all having the same negative charge and mass, which move at distances from the nucleus that are very great compared to the dimensions of the nucleus or of the electrons themselves. In this picture we at once see a striking resemblance to a planetary system, such as we have in our own solar system. Just as the simplicity of the laws that govern the motions of the solar system is intimately connected with the circumstance that the dimensions of the moving bodies are small in relation to the orbits, so the corresponding relations in atomic structure provide us with an explanation of an essential feature of natural phenomena in so far as these depend on the properties of the elements. It makes clear at once that these properties can be divided into two sharply distinguished classes.

To the first class belong most of the ordinary physical and chemical properties of substances, such as their state of aggregation, colour, and chemical reactivity. These properties depend on the motion of the electron system and the way in which this motion changes under the influence of different external actions. On account of the large mass of the nucleus relative to that of the electrons and its smallness in comparison to the electron orbits, the electronic motion will depend only to a very small extent on the nuclear mass, and will be determined to a close approximation solely by the total electrical charge of the nucleus. Especially the inner structure of the nucleus and the way in which the charges and masses are distributed among its separate particles will have a vanishingly small influence on the motion of the electron system surrounding the nucleus. On the other hand, the structure of the nucleus will be responsible for the second class of properties that are shown in the radioactivity of substances. In the radioactive processes we meet with an explosion of the nucleus, whereby positive or negative particles, the so-called α- and β-particles, are expelled with very great velocities.

Our conceptions of atomic structure afford us, therefore, an immediate explanation of the complete lack of interdependence between the two classes of properties, which is most strikingly shown in the existence of substances which have to an extraordinarily close approximation the same ordinary physical and chemical properties, even though the atomic weights are not the same, and the radioactive properties are completely different. Such substances, of the existence of which the first evidence was found in the work of Soddy and other investigators on the chemical properties of the radioactive elements, are called isotopes, with reference to the classification of the elements according to ordinary physical and chemical properties. It is

[1] Lecture delivered at Stockholm, December 11, 1922, on the occasion of the receipt of the Nobel prize in physics for the year 1922. English translation by Dr. Frank C. Hoyt.

Professor Niels Bohr was a theoretical nuclear physicist at Copenhagen University. He created the Bohr atomic model. In 1922, he was awarded the Nobel Prize in Physics "for his services in the investigation of the structure of atoms and of the radiation emanating from them" (The Great Republic, Washington, D.C.).

Chapter 1. Uranium and Plutonium

incredible event that has ever happened to me in my life. It was almost as incredible as if you fired a fifteen inch shell at a piece of tissue paper and it came back and hit you."

In 1919, Rutherford bombarded nitrogen with alpha particles. The experiments led him to the model of the atom, with a very small charged nucleus surrounded by orbiting electrons. Rutherford also became the first person to transmute one element into another element when he converted nitrogen into oxygen through a nuclear reaction.

Working with Niels Bohr (who discovered that electrons move in specific orbits), Rutherford theorized about the existence of neutrons, which could somehow compensate for the repelling effect of the positive charges of protons by causing an attractive nuclear force and preventing the nuclei from breaking apart.

In 1932, James Chadwick discovered the subatomic particle "neutron." That same year, John Cockcroft and Ernest Walton designed an electric circuit that generated a high DC voltage from a low-voltage AC or pulsing DC input. They used this circuit design to power their particle accelerator, performing the "splitting of the atom."

On December 17, 1938, in Berlin, Otto Hahn, Fritz Strassmann, and Lisa Meitner discovered that it was possible to split a uranium atom into two pieces by bombarding it with neutrons. However, the combined weight of these two pieces would be less than the original piece. They realized that a split uranium nucleus loses a small amount of mass, but this mass corresponds to an enormous amount of energy. The fission process often produces gamma photons and releases a very large amount of energy even by the energetic standards of radioactive decay. This result verified what Albert Einstein had proposed in 1905 with his famous equation: $E = mc^2$ (where "E" is energy, "m" is mass, and "c" is the speed of light in a vacuum).

After the discovery of fission, it did not take long for scientists to verify that a large nucleus of uranium fissions by splitting into two smaller nuclei, along with a few neutrons, accompanied by the release of heat energy and gamma rays. When the released neutrons produce fission in another nucleus, this is called a "chain reaction."

Niels Bohr brought the results of this experiment to the United States in 1939. When news of a fission breakthrough was verified, many scientists and political figures realized the enormous potential use and/or danger posed by atomic energy. Many believed that Nazi Germany would soon be in a position to harness this new resource.

Leo Szilard was a Hungarian-German-American physicist and inventor. He conceived the nuclear chain reaction in 1933. In 1934, he patented the idea of a nuclear fission reactor. In 1938, Szilard worked with Enrico Fermi and Walter Zinn on creating a nuclear chain reaction.

Chapter 2

The Manhattan Project

In late July 1939, Leo Szilard composed a letter to President Franklin Roosevelt and asked his friend Albert Einstein to sign it. The letter, dated August 2, 1939, was to be hand-delivered by Alexander Sachs, an economist, financier, and friend of the president.

Ten weeks later, on October 11, 1939, Sachs met with President Roosevelt.

Physicist Leo Szilard (right) and Albert Einstein discuss the letter to President Roosevelt. Szilard worked with the Enrico Fermi to achieve the initial self-sustained chain reaction of nuclear energy at the University of Chicago on December 2, 1942, which led to the development of the atomic bomb (Leo Szilard Papers, Special Collections & Archives, UC San Diego).

Chapter 2. The Manhattan Project

The Einstein-Szilard letter was a message written by Leo Szilard and signed by Albert Einstein that was sent to President Franklin D. Roosevelt on August 2, 1939. The letter warned that Germany might develop atomic bombs and suggested that the United States should start its own nuclear program. This proposal eventually led to the creation of the Manhattan Project. Autographed by Morris Jeppson, Tom Ferebee, Wyatt Duzenbury, Theodore Van Kirk, and Paul Tibbets (author's collection).

Rather than handing the Einstein letter to President Roosevelt, Sachs read the letter aloud, in order to ensure that the president would thoroughly appreciate the contents of the document. In addition to the message and an appended memorandum by Szilard, a much more comprehensive statement was presented to the president.

Einstein's letter explained to the president that there was recent research on fission chain reactions utilizing uranium. This development made it probable that large amounts of energy could be produced by a chain reaction. Furthermore, by harnessing this power, the construction of extremely powerful bombs was conceivable.

President Roosevelt replied to Einstein on October 19, 1939, informing him that a committee consisting of civilians and military representatives would study the potential uses of uranium.

Hotel King's Crown
420 West 116th Street
New York City

November 10, 1939

Dr. Alexander Sachs
One South William Street
New York City

Dear Dr. Sachs:

Dr. Fermi and I discussed the question which you have raised. It seems to us that it will be useful to have a small group of physicists, whose residence is not too far from Washington, D.C., consult with each other at regular intervals on questions connected with research on uranium. We attempted to draw up a list of names for this purpose. In our opinion such a list ought to include the following names to which others might be added, if required.

    Beams - Charlottesville, Va.
    Fermi - New York
    Furry - Cambridge, Mass.
    Szilard - New York
    Teller - Washington, D.C.
    Tuve - Washington, D.C.
    Wheeler - Princeton.

In drawing up this list we kept in mind two points:

a) the question of residence of the man selected. The geographical boundary line was drawn at the distance Washington to Boston;

b) the advisability of having a number of the more important eastern universities represented, at which research on uranium has been carried on in the past or might be started in the near future.

On November 10, 1939, Leo Szilard wrote a letter to Alexander Sachs (advisor to President Roosevelt) about forming a group to discuss research on uranium (Franklin D. Roosevelt Presidential Library and Museum).

This proved to be the first step that led to the creation of the Manhattan Project.

President Roosevelt appointed Lyman J. Briggs, director of the National Bureau of Standards, as head of the Advisory Committee on Uranium. The committee met two days later, on October 21, 1939, to coordinate its activities with Alexander Sachs, to look into all current research on uranium. On November 10,

-2-

Furthermore it seems to us that it might be useful to ask certain small groups of workers to consider themselves responsible for clearing up a given aspect of the question and to submit a report within six months' or a year's time. It would be the task of these groups to see to it that the questions involved are vigorously pursued, either by some members of the group or by others. Such a group would be expected to report at once if they encounter difficulties which they are unable to overcome, so that the help of others can be enlisted.

In our opinion the following persons might be asked to report on, and concern themselves with:

1. Slow Neutron Reaction: *all (Columbia)*
   Fermi, Pegram, Szilard, Wheeler *(Princeton)*
2. Fast Neutron Reaction:
   Fermi, Szilard, Tuve *(Carnegie Inst. Wash.)*, Wigner *(Princeton)*
3. The Question which of the Uranium Isotopes splits:
   Dunning *(Columbia)*, Fermi, Tuve, Wheeler.
4. Small Scale Separation of Isotopes by any Method except Diffusion:
   Beams *(Univ. Va)*, Fermi, Tuve.
5. Small Scale Separation of Isotopes by Diffusion:
   Fermi, Furry *(Harvard)*, Urey *(Columbia)*
6. Theoretical Possibility and Limitation of Large Scale Separation by Centrifuging:
   Beams, Pegram, Szilard, Teller *(George Washn. Univ.)*

List of potential scientists that are working on uranium (Franklin D. Roosevelt Presidential Library and Museum).

1939, Szilard wrote to Sachs, suggesting names of scientists that should be consulted at regular intervals on questions connected with uranium research.

In 1940, the Advisory Committee on Uranium started funding research by Enrico Fermi and Leo Szilard at Columbia University, which was focused on radioactive isotope separation (also known as uranium enrichment) and nuclear chain reactions.

The Advisory Committee on Uranium changed its name to the National Defense Research Committee in 1940. In 1941, it became the Office of Scientific Research and Development (OSRD).

On September 27, 1940, an agreement between Nazi Germany, Fascist Italy and Japan was signed in Berlin. In this Tripartite Pact (also known as the Berlin

```
            -3-

    7. Theoretical Possibility and Limitation of Large Scale
       Separation by Diffusion: (Yale)
           Fermi, Furry, Onsager, Urey.
    8. Possibility of Large Scale Production of Uranium Metal:
           Pegram, Szilard, and somebody from the Department for
           Chemical Engineering of MIT or Columbia.

       These groups include the following names:
       Beams    - University of Virginia,
       Fermi    - Columbia,
       Furry    - Harvard,
       Dunning  - Columbia,
       Pegram   - Columbia,
       Onsager  - Yale,
       Szilard  - Columbia,
       Teller   - George Washington University,
       Tuve     - Carnegie Institute of Terrestrial Magnetism,
       Urey     - Columbia,
       Wheeler  - Princeton,
       Wigner   - Princeton.

       We could not discuss the tentative proposals contained in this
    letter with Professor Pegram on account of his absence, and the time
    was too short to discuss it with anybody else.

                                        Yours very sincerely
                                        (Leo Szilard)
```

Detailed list with universities (Franklin D. Roosevelt Presidential Library and Museum).

Pact), each country agreed to assist the others should any of them be attacked by a country not already involved in the war.

Between September 22, 1940, and September 26, 1940, the Japanese invaded northern French Indochina in an effort to embargo all imports into China, including war supplies purchased from the United States. This move prompted the United States to embargo all oil exports, leading the Imperial Japanese Navy to estimate that it had less than two years of bunker oil remaining and to support the secret plans to seize oil resources in the Dutch East Indies.

On December 7, 1941, Japanese planes attacked the U.S. naval base on the island of Oahu, Hawaii. The attack was one of the greatest military surprises in the history of warfare. It took only two hours for the U.S. Pacific Fleet to be demolished. The attack destroyed or crippled 18 ships and killed 2,335 soldiers, sailors and marines.

Around 1:30 p.m. Eastern time, while the attack was under way, Secretary of the Navy Frank Knox told President Roosevelt that the Japanese had attacked Pearl Harbor. After meeting with his military advisors, President Roosevelt calmly and decisively dictated a letter to his secretary, Grace Tully. This letter called on Congress for a declaration of war.

Chapter 2. The Manhattan Project

TO THE CONGRESS OF THE UNITED STATES:

Yesterday, December 7, 1941 -- a date which will live in infamy -- the United States of America was suddenly and deliberately attacked by naval and air forces of the Empire of Japan.

The United States was at peace with that nation and, at the solicitation of Japan, was still in conversation with its Government and its Emperor looking toward the maintenance of peace in the Pacific. Indeed, one hour after Japanese air squadrons had commenced bombing in Oahu, the Japanese Ambassador to the United States and his colleague delivered to the Secretary of State a formal reply to a recent American message. While this reply stated that it seemed useless to continue the existing diplomatic negotiations, it contained no threat or hint of war or armed attack.

It will be recorded that the distance of Hawaii from Japan makes it obvious that the attack was deliberately planned many days or even weeks ago. During the intervening time the Japanese Government has deliberately sought to deceive the United States by false statements and expressions of hope for continued peace.

The attack yesterday on the Hawaiian Islands has caused severe damage to American naval and military forces. Very many American lives have been lost. In addition American ships have been reported torpedoed on the high seas between San Francisco and Honolulu.

The "Day of Infamy" speech was delivered by President Franklin D. Roosevelt to a joint session of the U.S. Congress on December 8, 1941, one day after the Empire of Japan's attack on the U.S. naval base at Pearl Harbor, Hawaii, and the Japanese declaration of war on the United States and the British Empire. This three-page letter was the first draft of a seven-minute speech. The name is derived from the first line of the speech: Roosevelt describing the previous day as "a date which will live in infamy" (National Archives and Records Administration).

- 2 -

Yesterday the Japanese Government also launched an attack against Malaya.

Last night Japanese forces attacked Hong Kong.

Last night Japanese forces attacked Guam.

Last night Japanese forces attacked the Philippine Islands.

Last night the Japanese attacked Wake Island.

This morning the Japanese attacked Midway Island.

Japan has, therefore, undertaken a surprise offensive extending throughout the Pacific area. The facts of yesterday speak for themselves. The people of the United States have already formed their opinions and well understand the implications to the very life and safety of our nation.

As Commander-in-Chief of the Army and Navy I have directed that all measures be taken for our defense.

Always will be remembered the character of the onslaught against us.

No matter how long it may take us to overcome this premeditated invasion, the American people will in their righteous might win through to absolute victory.

I believe I interpret the will of the Congress and of the people when I assert that we will not only defend ourselves to the uttermost but will make very certain that this form of treachery shall never endanger us again.

Hostilities exist. There is no blinking at the fact that our people, our territory and our interests are in grave danger.

President Roosevelt's speech was not intended merely as a personal response by the president but also as a statement on behalf of the entire American people. He expressed, in a few words, how the attack was deliberately planned by the Japanese. This speech helped the American people to work for a collective and decisive response (National Archives and Records Administration).

- 3 -

    With confidence in our armed forces -- with the unbounding determination of our people -- we will gain the inevitable triumph -- so help us God.

    I ask that the Congress declare that since the unprovoked and dastardly attack by Japan on Sunday, December seventh, a state of war has existed between the United States and the Japanese Empire.

Franklin D. Roosevelt

THE WHITE HOUSE,
    December 8, 1941.

Within an hour of Roosevelt's speech, Congress passed a formal declaration of war against Japan and officially brought the United States into World War II. This address is considered one of the most famous of all American political speeches (National Archives and Records Administration).

On December 8, at 12:30 p.m., the president addressed a joint session of Congress and, via radio, the nation. Delivering his famous "Day of Infamy Speech," President Roosevelt declared that the United States would enter World War II against the Japanese in the Pacific theater; when Germany declared war on the United States three days later, U.S. forces aligned with Great Britain, France and Russia to fight the Germans in Europe as well.

With President Roosevelt's approval, the Army Corps of Engineers joined the OSRD in 1942. It soon became a military initiative with scientists serving in a supportive role. The OSRD formed the Manhattan Engineer District on August 13, 1942, with headquarters in the New York borough of Manhattan. U.S. Army Colonel Leslie R. Groves was selected to lead the project. Groves was an outstanding West Point engineering graduate who had helped build the Pentagon.

This photo of J. Robert Oppenheimer, director of the Manhattan Project, was taken on September 18, 1945 (© 2010 The Regents of the University of California, Lawrence Berkeley National Laboratory).

On December 28, 1942, President Roosevelt authorized the creation of the Manhattan Project to combine the various research efforts with the goal of weaponizing nuclear energy. The cost would be over $2 billion, without the knowledge of the general public. Funds for projects all over the United States, involving over 100,000 men and women, would be hidden in the federal budget.

Facilities were set up in remote locations in New Mexico, Tennessee and Washington, in addition to sites in Canada, for this research and related atomic tests to be performed.

In 1941, at the University of California, Berkeley, Glenn T. Seaborg identified a new element—element 94 (plutonium). In May of the same year, Seaborg proved that plutonium (Pu-239) was 1.7 times as fissionable as uranium-235 (U-235), and thus a better nuclear explosive.

On August 20, 1942, Seaborg (now at the Metallurgical Laboratory, University of Chicago) identified a sequence of chemical oxidation and reduction cycles that produced a microgram (one millionth of a gram) of plutonium. He was quoted as saying, "Perhaps today was the most exciting and thrilling day I have

OFFICE OF WAR MOBILIZATION AND RECONVERSION
WASHINGTON, D. C.

OFFICE OF THE DIRECTOR

March 3, 1945.

MEMORANDUM FOR THE PRESIDENT

FROM: James F. Byrnes

I understand that the expenditures for the Manhattan project are approaching 2 billion dollars with no definite assurance yet of production.

We have succeeded to date in obtaining the cooperation of Congressional Committees in secret hearings. Perhaps we can continue to do so while the war lasts.

However, if the project proves a failure, it will then be subjected to relentless investigation and criticism.

I know little of the project except that it is supported by eminent scientists. Even eminent scientists may favor continuance of a project rather than to concede its failure. Also, it may be feasible to continue the experiment on a reduced scale.

In any event, no harm could come from an impartial investigation and review by a small group of scientists not already identified with the project.

Such a review might hurt the feelings of those now engaged in the project. Still 2 billion dollars is enough money to risk such hurt.

A favorable finding would justify continuance. An unfavorable finding would at least indicate the need for further justification by those who are responsible.

In any event, it would be clear that we were mindful of the tremendous expenditure of men and materials.

In this letter, James F. Byrnes warns President Roosevelt that the Manhattan Project cost is almost $2 billion. During the war, Byrnes led the Office of Economic Stabilization and the Office of War Mobilization and was a candidate to replace Henry A. Wallace as Roosevelt's running mate in the 1944 election; however, in the end, Harry S. Truman was nominated by the Democratic National Convention (Franklin D. Roosevelt Presidential Library and Museum).

experienced. Our microchemists isolated pure element 94 (plutonium) for the first time.... It is the first time that element 94 has been beheld by the eye of man."

Enrico Fermi led the team that designed and built Chicago Pile-1 (the world's first artificial nuclear reactor) under the abandoned west stands of the University of Chicago's Stagg Field. It used a stack of 40,000 graphite blocks, held together in a wooden frame, 25 feet wide and 20 feet tall. Half of the blocks contained uranium oxide, while pure graphite rods were inserted in the pile to control the reaction. These rods would absorb the neutrons.

The world's first manmade nuclear chain reaction occurred on December 2, 1942. It was shut down a half-hour later, before its growing heat and radioactivity became too dangerous. However, Fermi and Szilard had successfully enriched uranium-238 to produce U-235. (Author's note: By capturing a neutron, U-238 becomes U-239, which rapidly changes by beta radiation into neptunium-239. This neptunium is transformed by beta radiation, after 3 days average, into a new nucleus: P-239.)

Reunion of atomic scientists on the fourth anniversary (1946) of the first controlled nuclear fission chain reaction on December 2, 1942. Pictured in front of Bernard A. Eckhart Hall at the University of Chicago, from left (third row): Norman Hilberry, Samuel Allison, Thomas Brill, Robert G. Nobles, Warren Nyer, Marvin Wilkening; (second row): Harold Agnew, William Sturm, Harold Lichtenberger, Leona W. Marshall, Leo Szilard; (first row): Enrico Fermi, Walter H. Zinn, Albert Wattenberg, Herbert L. Anderson (Special Collections Research Center, University of Chicago Library).

Opposite, top: Glenn T. Seaborg discovered ten transuranium elements, which earned him a share of the 1951 Nobel Prize in Chemistry. Seaborg joined the chemistry group at the Metallurgical Laboratory of the Manhattan Project at the University of Chicago, where Enrico Fermi and his group would later convert uranium-238 to plutonium-239 in a controlled nuclear chain reaction. Seaborg's role was to figure out how to extract the tiny bit of plutonium from the mass of uranium (AIP Emilio Segrè Visual Archives, Physics Today Collection, University of Chicago Photographic Archive, apf1-07530, Special Collections Research Center, University of Chicago Library). *Opposite, bottom:* Major General Leslie R. Groves, director of the Manhattan District, pins a Medal of Merit on physicist Enrico Fermi for his contribution to the success of the atomic bomb project. They are pictured at the University of Chicago with other scientists who also received the award. *From left:* General Groves, Harold C. Urey (Martin A. Ryerson Distinguished Professor of Chemistry), Dr. Fermi, Samuel K. Allison (director of the university's Institute for Nuclear Studies), Cyril S. Smith (director of the university's Institute for the Study of Metals), and Robert S. Stone of the University of California Hospitals. Photograph taken March 20, 1946 (University of Chicago Library, Special Collections Research Center).

The Metallurgical Laboratory experimented with Pile-1 for several months and then physically moved it to a new location, where it became known as Chicago Pile-2. Data from this pile would later be used for the design of reactors, including the one that furnished the plutonium for the Trinity and Nagasaki atomic bombs.

Chapter 3

Los Alamos National Laboratory

The story of the Manhattan Project and its creation of the atomic bomb is one of the greatest scientific achievements of all times. The founding of the Los Alamos National Laboratory set the stage for its successful secret development of the atomic bomb.

The United States Army purchased a site in the Pajarito Plateau at Los Alamos, New Mexico, on November 25, 1942. It consisted of approximately 54,000 acres and 27 houses, dormitories, and miscellaneous buildings.

On January 1, 1943, the University of California was selected to operate a new laboratory, the Los Alamos Laboratory (code named "Project Y"). Project Y

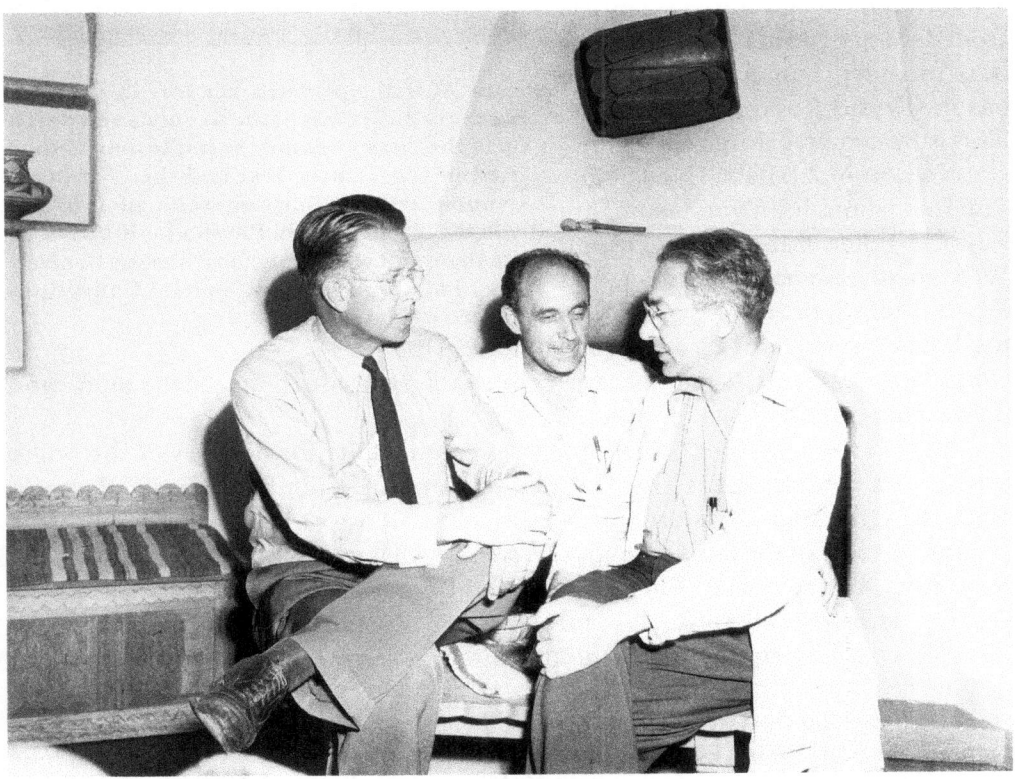

Manhattan Project physicists, from left: Ernest O. Lawrence, Enrico Fermi, and Isidor I. Rabi (University of Chicago Library, Special Collections Research Center).

was established to design and build atomic bombs.

J. Robert Oppenheimer, who was already working on the concept of nuclear fission (with Edward Teller), was named director of the new laboratory. Groves selected Oppenheimer despite the fact that Oppenheimer's closest friends were active members in the Communist Party. Oppenheimer was in favor of using the atomic bomb to put a swift end to the war.

Oppenheimer spent the first few months of 1943 crisscrossing the United States and gathering the best scientists and engineers for this mammoth project. He also brought in acclaimed scientists from Great Britain. Project members included Luis Alvarez, Enrico Fermi, Bruno Rossi, Emilio Segre, Norris Bradbury, George Kistiakowsky, Niels Bohr, I.I. Rabi, Hans Bethe, Rolf Landshoff, John von Neumann, Edward Teller, Otto Frisch, Joseph W. Kennedy, Richard Feynman, and Edwin McMillan, among others. In total, there were about 2,700 people, consisting of scientists, engineers, technicians and assorted security staff, who worked in secret to develop the bomb.

Hans A. Bethe played a key role in calculating the critical mass of the weapons and developed the theory behind the implosion method used in the Trinity Test and the "Fat Man" weapon dropped on Nagasaki. In 1967, he won the Nobel Prize in Physics for his work on the theory of stellar nucleosynthesis (University of Chicago Library, Special Collections Research Center).

Within the Los Alamos Laboratory, responsibility for delivery of the bomb lay with its Ordnance Division, headed by Captain William S. "Deak" Parsons. With the Ordnance Division, the E-7 Group was created with responsibility for the integration of design and delivery. Led by physicist Norman F. Ramsey, the team consisted of himself, Sheldon Dike and Bernard Waldman.

Oppenheimer advocated a central facility where theoretical and experimental work could be conducted without military interference. He proved to be an

Opposite, bottom: **President Kennedy pays a visit to Lawrence Berkeley Laboratory. Emerging from Building 70A, left to right, are Norris Bradbury (LASL director), John Foster (LRL Livermore director), Edwin McMillan (LRL director), Glenn Seaborg (AEC chairman), the president, Edward Teller (LRL associate director), Robert McNamara (defense secretary), and Harold Brown (director of defense research and engineering).** *Magnet* **6, no. 4 (April 1962), 4–5. Laboratory Directorate (© 2010 The Regents of the University of California, Lawrence Berkeley National Laboratory).**

Left: Robert F. Bacher was an American nuclear physicist. At Los Alamos, Bacher was head of the Manhattan Project's Physics Division and, later, its Gadget Division. Autographed by Robert Bacher (Los Alamos National Laboratory; author's collection). *Right:* Edward Teller was an American nuclear physicist who participated in the production of the first atomic bomb (1945) and led the development of the world's first thermonuclear weapon (the hydrogen bomb). As a member of the Manhattan Project, Teller proposed the successful solid-pit implosion design. Autographed by Edward Teller (U.S. Department of Energy; author's collection).

Otto R. Frisch was an Austrian-born British physicist who worked on nuclear physics. With Lise Meitner, he advanced the first theoretical explanation of nuclear fission (coining the term) and first experimentally detected the fission byproducts. While working for the Manhattan Project, Frisch calculated the size of the critical mass of U-235 needed for an atomic bomb. He also predicted the effects of the blast and fallout of an explosion. Autographed by Otto Frisch (author's collection).

excellent director, satisfying the emotional and intellectual needs of his highly distinguished staff.

In a short period of time, the Los Alamos Laboratory acquired particle accelerators and other experimental equipment, including two Van de Graaff generators, a Cockcroft-Walton machine, and a cyclotron.

In addition to all the calculations on uranium and plutonium fission, chain reactions, and critical masses, scientists worked on the ordnance aspects of the bomb. The "Gadget," as it became to be known, would require two subcritical masses of fissionable material to come together to form a supercritical mass for an explosion to occur.

For the uranium bomb, a conventional artillery method of firing one subcritical mass into another was chosen. This gun-type nuclear device relies on one subcritical mass of a fissile material being fired into another subcritical mass with the use of high explosives such that the assembly becomes supercritical. The explosive must force the additional mass into contact for about a microsecond—the time it takes for the rapid chain reaction to produce the nuclear explosion. This was the design of the Hiroshima bomb, which used 64 kilograms of U-235 and produced an explosion equivalent in force to 15 kilotons of trinitrotoluene (TNT).

In the case of plutonium, which was much more difficult, an "implosion method" was used. Here, one had to fashion a subcritical mass of Pu-239 into

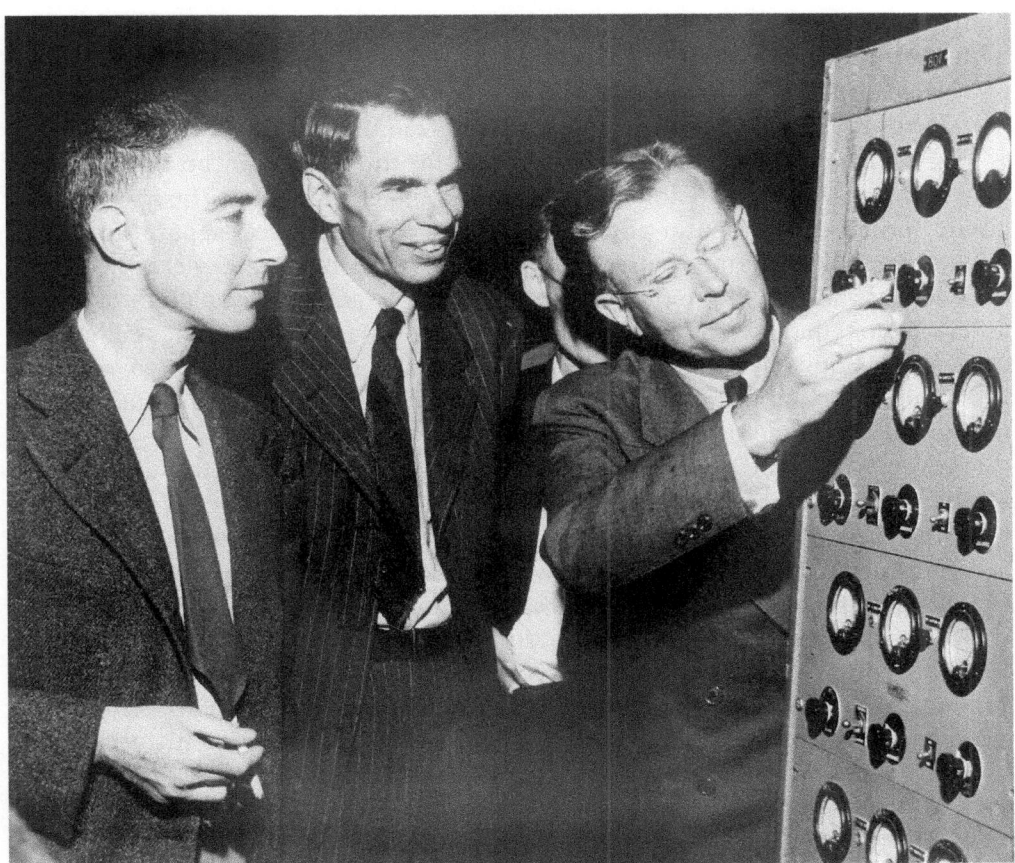

Ernest O. Lawrence, Glenn T. Seaborg, and J. Robert Oppenheimer in early 1946 at the controls to the magnet of the 184-inch cyclotron, which was being converted from its wartime use to its original purpose as a cyclotron (© 2010 The Regents of the University of California, Lawrence Berkeley National Laboratory).

a spherical shape and then set off high explosives to drive it inward. The spherical shell of high-explosive material is made up of lenses that focus the explosion inward. The explosives must be detonated simultaneously. This process creates enough force to increase the density of the sphere of plutonium to the point that it becomes supercritical. This was the design of the Trinity and Nagasaki bombs. The Nagasaki bomb used 6.4 kilograms of plutonium and produced on explosion equivalent to 21 kilotons of TNT.

Enrico Fermi took control of the critical mass experiments and standardization of measurement techniques. Under Groves' direction, the engineering division was divided into four subdivisions: theoretical (Hans A. Bethe); experimental physics (Robert F. Bacher); chemistry and metallurgy (Joseph W. Kennedy); and ordnance (Navy Captain William Parsons).

The first major hurdle was finding enough uranium to make a bomb. Natural mined uranium is 99 percent U-238. It is a heavy isotope unfit for use in a bomb since it has a tendency to capture neutrons without fission. U-235, the lighter isotope needed for building a bomb, is seven-tenths of 1 percent of the natural-occurring uranium.

Separating U-235 from U-238 became the first obstacle. In addition, the metallurgy division had to turn the purified U-235 into metal.

In February 1943, DuPont broke ground for the X-10 "Clinton Pile Graphite Reactor" (also called the "Clinton Pile") at Oak Ridge, Tennessee. The site included an air-cooled experimental pile reactor and a pilot chemical separation plant. After several different methods for separating U-238 from U-235 were unsuccessful, a new separation method was discovered. The scientists used a method known as "electromagnetic separation." Electromagnetic separating machines called "calutrons," developed by the University of California, were installed. On November 4, 1943, the X-10 complex went critical for the first time. Within a few months, it was producing small amounts of experimental plutonium samples.

Invented by Ernest Lawrence during the Manhattan Project, the calutron was a mass spectrometer designed and used for separating the isotopes of uranium. At Oak Ridge, enough uranium was produced for "Little Boy" (the Hiroshima bomb).

On April 5, 1944, Emilio Segrè, who was also working on plutonium fission at Los Alamos, received the first sample of reactor-produced plutonium from Oak Ridge. In the same year, construction of a nuclear reactor began in Hanford, Washington, where water from the Columbia River could be used as a reactor coolant. By September 1944, the first Hanford pile was in operation.

The laboratory received small quantities of plutonium (as nitrate, not as metal) in October 1943. Larger amounts would follow, first from Oak Ridge, and then from Hanford.

In the spring of 1944, the world's first piece of solid plutonium metal was produced by the graphite centrifuge method. This method used centrifugal force to throw down molten metal into the tip of a cone during reduction.

The choice of a gun-type plutonium bomb became impractical when Segrè determined that reactor-bred plutonium had a higher concentration of the isotope plutonium-240 than cyclotron-produced plutonium. As plutonium-240 has a higher spontaneous fission rate, the original gun-type plutonium weapon (code named "Thin Man") was shelved. Instead, an implosion design would use a series of explosive lenses to compress a solid sphere of plutonium-239 into a high-density core, initiating a nuclear chain reaction.

The components for building the bomb were manufactured in three different plants, so that no one would have a copy of the complete design.

In March 1945, the B reactor at Hanford began producing plutonium for the implosion-type atomic bomb. Plutonium from the Hanford reactors ultimately fueled the detonation near Alamogordo, New Mexico, on July 16, 1945 (the Trinity Test), and the bomb ("Fat Man") that was dropped over Nagasaki, Japan, on August 9, 1945.

The designers for the "Little Boy" bomb were certain that it would function properly. However, many had misgivings about "Fat Man." To make certain "Fat Man" would detonate properly, the decision was made to make a test at Alamogordo, New Mexico.

Chapter 4

The First Atomic Bomb: Trinity

Besides the concerns over the performance of "Fat Man," it was essential to obtain quantitative analysis of the various effects of the new weapon. These included the destructive effects of heat, blast, and shock waves, as well as radiation intensities, fallout and the size of the fireball. All these and other measurements could not be obtained with complete accuracy in combat situations.

Oppenheimer chose the code name "Trinity" for the test. He assigned Samuel Allison, Robert Bacher, and George Kistiakowsky to take charge of the implosion weapon. Kenneth T. Bainbridge was put in charge of Project Trinity—a new division to oversee the July nuclear test.

As head of the Manhattan Project's Trinity nuclear test, Bainbridge, a physicist from Harvard, was responsible for preparations for a field test in which blast, earth shock, and neutron and gamma radiation would be studied. Also, a complete photographic record of the explosion would be obtained.

Captain William "Deak" Parsons was assigned to head Project Alberta (also known as Project A), which had the task of preparing and delivering the first nuclear weapons. Project Alberta consisted of 51 U.S. Army, Navy, and civilian personnel, including one British scientist. Its mission was three-fold: it had to design a bomb shape for delivery by air and then procure and assemble it. It supported the ballistic testing work at Wendover Army Airfield, Utah, conducted by the 216th Army Air Forces Base Unit (Project W-47), and the modification of B-29s to carry the bombs (Project Silverplate).

As head of Project Alberta, Parson appointed Norman Ramsey as his scientific and technical deputy and Frederick Ashworth as his operations officer and military alternate. There were two bomb assembly teams: a Fat Man Assembly Team under Commander Norris Bradbury and Roger Warner, and a Little Boy Assembly Team under Francis Birch.

Philip Morrison was the head of the pit crew. Bernard Waldman and Luis Alvarez led the Aerial Observation Team, while Sheldon Dike was in charge of the Aircraft Ordnance Team. Physicists Robert Serber and William Penney, as well as U.S. Army Captain James F. Nolan (a medical expert), were special consultants. All members of Project Alberta had volunteered for the mission.

On September 7, 1944, at a meeting with the commander of the Second Air

LOS ALAMOS
SANTA FE, NEW MEXICO
P. O. BOX 1663

July 9, 1943

Dear Mr. President:

Thank you for your generous letter of June 29th. You would be glad to know how greatly your good words of reassurance were appreciated by us. There will be many times in the months ahead when we shall remember them.

It is perhaps appropriate that I should in turn transmit to you the assurance that we as a group and as individual Americans are profoundly aware of our responsibility, for the security of our project as well as for its rapid and effective completion. It is a great source of encouragement to us that we have in this your support and understanding.

Very sincerely yours,

J. R. Oppenheimer

The President
The White House
Washington, D. C.

In this letter dated July 9, 1943, Oppenheimer thanks President Roosevelt for his good wishes (Franklin D. Roosevelt Presidential Library and Museum).

Force, it was decided that an 18-by-24-mile section of the northwest corner of the Alamogordo Bombing Range would be used as a test site.

Before the final test of Trinity, it was necessary to do a trial test—one that involved detonating 100 tons of conventional high explosives. This test would provide data for the calibration of instruments for the blast and shock measurements and would serve as a dress rehearsal for the real explosion.

Chapter 4. The First Atomic Bomb: Trinity

Scientific and military personnel on Tinian Island before the raids on Japan. *Bottom row, left to right*: Victor A. Miller, Arthur B. Machen, Roger S. Warner, Harlow W. Russ, Norman F. Ramsey, Edward B. Doll, John L. Tucker, George T. Reynolds, Milo Bolstad, W. R. Prohs. *Second row, left to right*: Charles P. Baker, Phillip Morrison, F. G. Penny, Theodore Perlman, Thomas H. Olmstead, A. Francis Birch, Admiral William R. Purnell, General Thomas Farrell, William S. Parsons, Frederick L. Ashworth, Robert Serber, Lawrence Langel, Bernard Waldman, Luis Alvarez, James F. Nolan. *Third row, left to right*: John D. Hopper, Edward G. Carlson, Leonard Motichko, Henry Linschitz, Benjamin B. Bederson, Raemer E. Schreiber, Walter Goodman, William L. Murphy, Robert P. Mathews, Robert W. Dawson, Harold M. Agnew, Lawrence H. Johnson, Edward Stephenson, David Anderson, Donald Mastick. *Top row, left to right*: Morton Camac, Jesse Kupferberg, E. J. O'Keefe, Eugene L. Nooker, Gunnar Thornton, William J. Larkin, Donald C. Harms, Frank J. Fortine, Frederick H. Zimmerli, Vincent Caleca, Arthur W. Collins. Autographed by Luis Alvarez (U.S. Army; author's collection)..

The trial took place on May 7, 1945, and proved to be a tremendously valuable experience. It provided information that would be essential to the success of the Trinity bomb test.

In early July 1945, J.M. Hubbard, who had joined the Trinity Project as a meteorologist, undertook the job of determining the best test date from a weather point of view.

On July 11, 1945, two hemispheres of plutonium made the trip to Trinity from Los Alamos. At 1:00 a.m. on Friday, July 13, 1945, the preassembled high-explosive components started for Trinity in a truck, convoyed by Army Intelligence, with George Kistiakowsky protecting the cargo.

In late afternoon, the active material and the high explosives came together for the first time. However, the assembly crew discovered that the two principal parts of the gadget, carefully designed and precision machined, no longer fit together.

The problem was quickly solved by Marshall Holloway, in charge of pit assembly. It turned out that the plutonium component had generated its own heat during the trip from Los Alamos and had expanded. A little wait time fixed the problem.

The next morning, the detonation crew, headed by Kenneth Greisen, climbed the 100-foot tower and made the final installations. Greisen returned every six hours to verify that the necessary cables were properly connected.

Late that night, the "Gadget" was ready, and an armed guard and scientists and technicians were left with a few routine preparations and last-minute adjustments on their equipment.

The Civil Aeronautics Authority, the Air Corps, the Navy and the Alamogordo base were barred from the entire area during the last important hours. No aircraft would be permitted until notification was given.

As with all high-profile events, official dignitaries were invited to the climatic demonstration. On Sunday afternoon, General Groves arrived with Vannevar Bush and James B. Conant, members of the policy committee for the Manhattan Engineer District (MED). Later a bus arrived with Charles A. Thomas, MED's coordinator for chemical research, along with Ernest O. Lawrence and Sir James Chadwick. Only one newsman was assigned by MED to document the development of the bomb: William L. Laurence from the *New York Times*. He would not report any of his observations and interviews with the crews and scientists until after the atomic bombing of Hiroshima.

In total, there are about 450 scientists and technicians at the Alamogordo test site.

Before Sunday, at 11:00 p.m., the arming party, consisting of Bainbridge, Kistiakowsky, and Joe McKibben, arrived with two Army weathermen. A total of 125 men under Lieutenant Howard Bush's command guarded the tower. McKibben arrived with a checklist of 47 jobs that had to be done before Zero Hour.

The next morning, July 16, 1945, at 2:00 a.m., Groves, monitoring the weather, decided to postpone the planned test from 4:00 a.m. to 5:30 a.m. The crucial weather report came at 4:45 a.m.: "Winds aloft very light, variable to

Chapter 4. The First Atomic Bomb: Trinity

Norris E. Bradbury was head of E-5 (the implosion experimentation group). Bradbury was in charge of the final assembly of the "Gadget" detonated on July 16, 1945. This bomb was placed on top of the 100-foot tower at Alamogordo, New Mexico (Los Alamos National Laboratory).

40,000, surface calm, Inversion about 17,000 ft., Humidity 12,000 to 18,000 above 80%, Conditions holding for next two hours, sky now broken, becoming scattered."

Bainbridge, McKibben and Kistiakowsky drove with Lieutenant Bush to the west 900-yard point, where, according to McKibben's checklist, he opened all "customer circuits." At the tower, connections were checked, switches were thrown, and arming, power, firing and informer leads were connected.

B-29s from Kirtland Army Air Field were supposed to fly over the explosion and drop gauges to collect information from the explosion. However, the weather in Albuquerque was terrible, and the decision was made to wait for the weather to improve. When the B-29s finally arrived, they were merely observers.

After 5:00 a.m., Bainbridge used his special key to unlock the lock that protected the switches from tampering while the arming party was at the tower.

At 5:10 a.m., Dr. Samuel Allison began his job as the "countdown man." Allison was director of the Metallurgical Lab at the University of Chicago during the Manhattan Project. He worked on the development of the Chicago Pile-1 alongside Leo Szilard, Eugene Wigner, and Enrico Fermi.

Everyone had been instructed to lie face down on the ground with their feet toward the blast and to close their eyes and cover them as the countdown approached zero. When they became aware of the flash, they were told that they could turn over and watch through the darkened glass that had been supplied to them.

Twenty miles away, at an observation post, Edward Teller and Fred Reines (a former Los Alamos physicist) waited with Kenneth Greisen and I.I. Rabi, a project consultant.

A few minutes before detonation, General Thomas Farrell would later recall his observations (recorded in the Memorandum for the Secretary of War on July 18, 1945):

> The scene inside the shelter was dramatic beyond words. In and around the shelter were some twenty-odd people concerned with last minute arrangements prior to firing the shot. Included were: Dr. Oppenheimer, the Director who had borne the great scientific burden of developing the weapon from the raw materials made in Tennessee and Washington and a dozen of his key assistants—Dr. Kistiakowsky, who developed the highly special explosives; Dr. Bainbridge, who supervised all the detailed arrangements for the test; Dr. Hubbard, the weather expert, and several others. Besides these, there were a handful of soldiers, two or three Army officers and one Naval officer. The shelter was cluttered with a great variety of instruments and radios.
>
> In that brief instant in the remote New Mexico desert the tremendous effort of the brains and brawn of all these people came suddenly and startlingly to the fullest fruition. Dr. Oppenheimer, on whom had rested a very heavy burden, grew tenser as the last seconds ticked off. He scarcely breathed. He held on to a post to steady himself.

Sam Allison continued the countdown until 5:29:45 a.m.

At first, an incredible burst of light filled the surrounding mountains with the brilliance of ten suns. A gigantic fireball erupted with a tremendous roar and started expanding away from "Ground Zero." The cloud radiated heat and started to form a purple-white mushroom, rising to 40,000 feet. Then there was a powerful shock wave that knocked over two military men two miles away.

The 100-foot steel tower that held the bomb at "Ground Zero" had disintegrated. It became gas and vanished. Temperature has been estimated to have been 100 million degrees Fahrenheit. The estimated yield of the bomb's explosive energy was 19 kilotons.

In 1967, the Los Alamos Scientific Laboratory published the complete story of the Manhattan Project: *Los Alamos: The Beginning of an Era*. In this fabulous book, Enrico Fermi is quoted as saying:

> My first impression of the explosion was the very intense flash of light, and a sensation of heat on the parts of my body that were exposed. Although I did not look directly towards the object, I had the impression that suddenly the countryside became brighter than in full daylight. I subsequently looked in the direction of the explosion through the dark glass and could see something that looked like a conglomeration of flames that promptly started rising. After a few seconds the rising flames lost their brightness and appeared as a huge pillar of smoke with an expanded head like a gigantic mushroom that rose rapidly beyond the clouds, probably to a height of the order of 30,000 feet. After reaching full height, the smoke stayed stationary for a while before the wind started dispersing it.

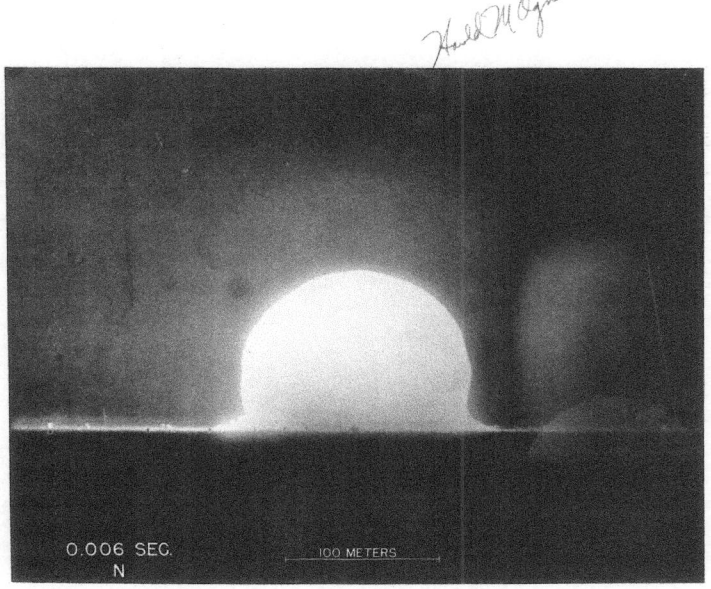

Trinity atomic bomb test, July 16, 1945, Alamogordo, New Mexico. This photograph taken from five miles away, 0.006 seconds after detonation. Photographs were taken by Berlyn Brixner, using a 35-mm Mitchell camera. Autographed by Harold M. Agnew (Agnew is best known for helping to create the first "chain-reacting pile" with Enrico Fermi at Columbia University and the University of Chicago. He worked on the Manhattan Project and was a scientific observer on the Hiroshima mission) (Los Alamos Photographic Laboratory; author's collection).

Photo taken 0.016 seconds after detonation by Berlyn Brixner. Autographed by Berlyn Brixner (Los Alamos Photographic Laboratory; author's collection).

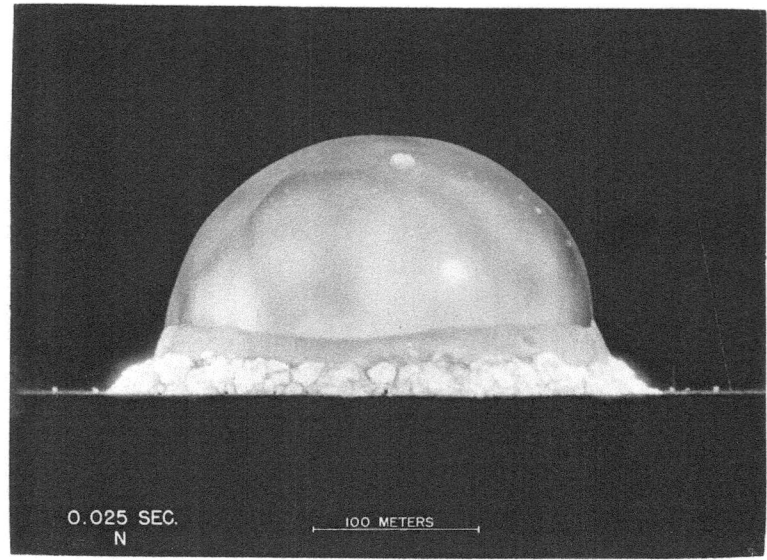

Photo taken 0.025 seconds after detonation by Berlyn Brixner. Autographed by Berlyn Brixner (Los Alamos Photographic Laboratory; author's collection).

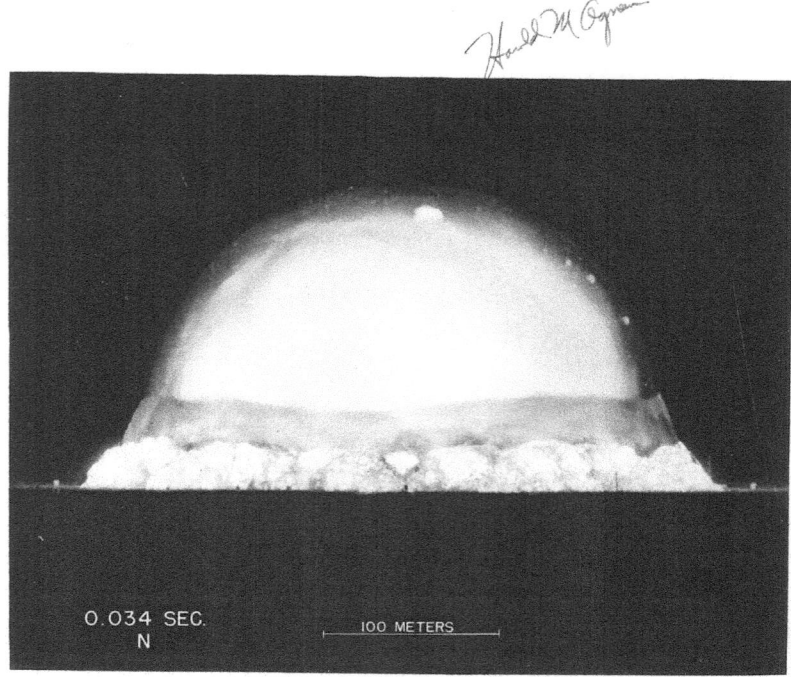

Photo taken 0.034 seconds after detonation by Berlyn Brixner. Autographed by Harold Agnew (Los Alamos Photographic Laboratory; author's collection).

Chapter 4. The First Atomic Bomb: Trinity

Photo taken 0.053 seconds after detonation by Berlyn Brixner. Autographed by Norris Bradbury and Robert Bacher (Los Alamos Photographic Laboratory; author's collection).

Photo taken 0.100 seconds after detonation by Berlyn Brixner. Autographed by Norris Bradbury and Robert Bacher (Los Alamos Photographic Laboratory; author's collection).

Photo taken 2.0 seconds after detonation by Berlyn Brixner. Autographed by Robert Bacher (Los Alamos Photographic Laboratory; author's collection).

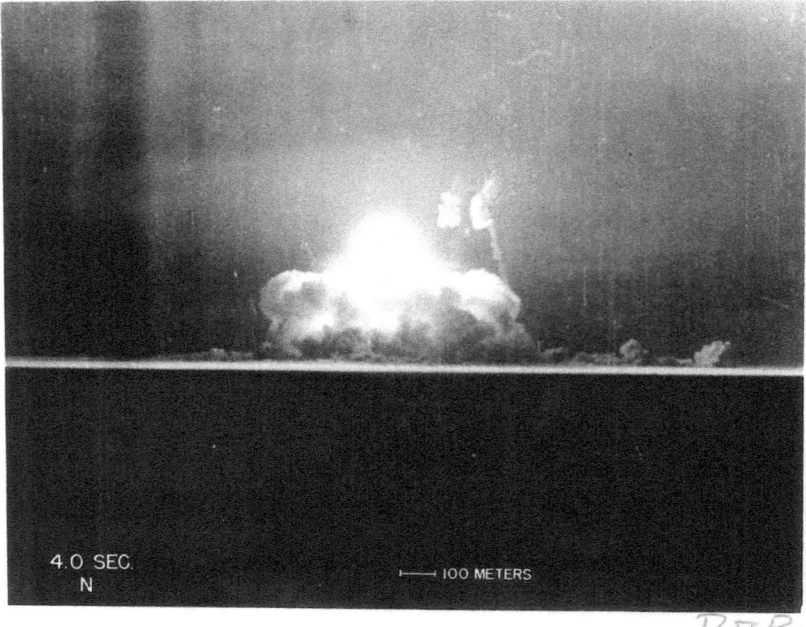

Photo taken 4.0 seconds after detonation by Berlyn Brixner. Autographed by Robert Bacher (Los Alamos Photographic Laboratory; author's collection).

Chapter 4. The First Atomic Bomb: Trinity

Photo taken 9.0 seconds after detonation by Berlyn Brixner. Autographed by Robert Bacher (Los Alamos Photographic Laboratory; author's collection).

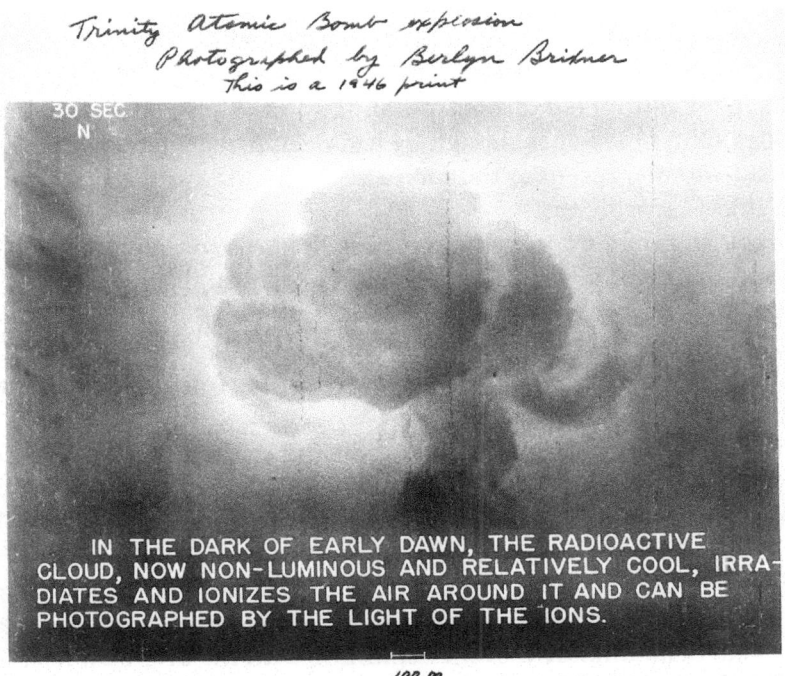

Photo taken 30.0 seconds after detonation by Berlyn Brixner. Autographed by Berlyn Brixner (Los Alamos Photographic Laboratory; author's collection).

In 1990, Robert Raymond wrote a fascinating book on how metals have changed our existence. In *Out of the Fiery Furnace*, he quotes General Farrell's observations on the Trinity test:

> The effects could well be called unprecedented, magnificent, beautiful, stupendous, and terrifying. No man-made phenomenon of such tremendous power had ever occurred before. The lighting effects beggared description. The whole country was lighted by a searing light with the intensity many times that of the midday sun. It was golden, purple, violet, gray and blue. It lighted every peak, crevasse and ridge of the nearby mountain range with a clarity and beauty that cannot be described but must be seen to be imagined. Seconds after the explosion came, first, the air blast pressing hard against the people, to be followed almost immediately by the strong, sustained awesome roar which warned of doomsday and made us feel we puny things were blasphemous to dare tamper with the forces heretofore reserved for the Almighty.

Dr. Hans Bethe, one of the Los Alamos scientists, is quoted in a booklet called *Trinity Site: 1945–1995*, issued by the White Sands Missile Range, Public Affairs Office: "It looked like a giant magnesium flare which kept on for what seemed a whole minute but was actually one or two seconds. The white ball grew and after a few seconds became clouded with dust whipped up by the explosion from the ground and rose and left behind a black trail of dust particles."

William L. Laurence, a science writer and the only reporter permitted to attend the test, would write his observations in the September 26, 1945, edition of the *New York Times*: "It was like the grand finale of a mighty symphony of the elements, fascinating and terrifying, uplifting and crushing, ominous, devastating, full of great promise and great foreboding. On that moment hung eternity. Time stood still. Space contracted to a pin-point. It was as though the earth had opened and the skies split. One felt as though he had been privileged to witness the Birth of the World—to be present at the moment of Creation when the Lord said: Let there be light."

The explosion had depressed an area 400 yards in diameter, and the sand had turned into green glass-like marbles. (These small pieces—still radioactive—would later be called Trinitite or Alamogordo glass.) The shock wave broke windows 120 miles away, and thousands of people (some as far away 150 miles) phoned the police with questions after hearing an explosion and seeing bright lights in the sky.

Julian Mack, an American physicist, was the leader of Group G-11 (Optics), the photography group, in the Weapons Physics Division at Los Alamos during the Manhattan Project. He coordinated taking high-speed photographs of the Trinity test.

As chief photographer for Mack, Berlyn Brixner was assigned to shoot movies in 16-millimeter black-and-white format, from every angle and distance and at every speed, of an unknown event beginning with the brightest flash ever produced on Earth.

He was told that the brightness of the "unknown event" would be approximately 10-sun brightness. Brixner later commented, "All I had to do was go out

Chapter 4. The First Atomic Bomb: Trinity

Robert Oppenheimer (left) and General Leslie Groves (right) at Ground Zero after the nuclear explosion. At their feet are the remains of the 100-foot steel tower (Digital Photo Archive, Department of Energy, courtesy AIP Emilio Segrè Visual Archives).

and point my camera at the sun and take some pictures. Ten times that was easy to calculate."

Brixner had 55 cameras set up at various distances from the detonation site and at various angles. Some cameras were 20 miles away. The two cameras closest to the tower were shielded with 12-inch-thick leaded glass.

> 1342 46th St.
> Los Alamos, NM 87544
> 1 May 1989
>
> Dear Mr. Kloda —
>
> Enclosed are two autographed photos (0.012 + 30. sec. 1946 glossy prints) that I took of the explosion at the Trinity atomic Bomb test. I was assigned the job of getting the motion-picture record and had many cameras set up to get good coverage. These photos are from frames of the 35-mm Mitchell camera films. The complete movie record was shown on national newsreels soon after VJ day.
>
> Sincerely,
>
> Berlyn Brixner

Berlyn Brixner's note describing what camera he used to film the Trinity test. Autographed by Berlyn Brixner (author's collection).

Craig Nelson, in *The Age of Radiance: The Epic Rise and Dramatic Fall of the Atomic Era*, quotes Brixner's observations: "I was temporarily blinded. I looked to the side. The Oscura mountains were as bright as day. I saw this tremendous ball of fire, and it was rising. I was just spellbound! I followed it as it rose. There was no sound! It all took place in absolute silence. Then it dawned on me. I'm the photographer! I've gotta get that ball of fire."

Thirty seconds after the atomic blast, most of Brixner's 55 cameras in the desert were finished. He would have over 100,000 frames to develop in black and white and a few in the new temperamental Kodachrome.

Weeks before the Trinity test, a young civilian member of the Special Engineering Detachment was assigned to help Italian physicist Emilio Segrè of the Los Alamos Theoretical Division. Segrè was studying delayed gamma rays. His technician was Jack Aeby.

Aeby helped set up radiation detectors near the detonation tower. These instruments were hung on barrage balloons tethered 800 yards from the tower. In a millisecond after they transmitted their nuclear data, they were all vaporized.

On the day of the Trinity test, Segrè secured permission for Aeby to carry his personal 35-millimeter still camera to the site to record the group's activities. When the countdown started, Aeby, at North 10,000, was planning to take a new Anscochrome color transparency picture of the bomb.

Before the detonation, Aeby donned his special goggles and put his Perfex 44 camera on "bulb." In the dark before "Zero," he opened up the shutter, figuring that way he'd get a good image of the flash.

After Aeby saw the white flash in his goggles, he removed them and reset his camera. In *The Age of Radiance*, Nelson quotes Aeby as to what followed: "I released the shutter, cranked the diaphragm down, changed the shutter speed and fired three times in succession. I quit at three because I was out of film."

Later that night, Aeby took his film—a nonstandard piece of Anscochrome movie stock film—out of his camera at a local photo lab. He worked it through the 21-step procedure for color film developing. Only one of the three developed shots was good.

In fact, Aeby had taken the only well-exposed color photograph of the first detonation of an atomic bomb at the Trinity test site in New Mexico. This photograph provided the basis for the Theoretical Division's leader, Edward Teller, to calculate the yield of the Trinity weapon.

Aeby made seven copies of his photograph. Later the Army asked for six of the seven prints, plus the original negative, "for safe keeping." The Army subsequently "lost" these items.

On July 16, 1945, the commanding officer of the Alamogordo Army Air Base made the following statement:

> A remotely located ammunition magazine containing a considerable amount of high explosives and pyrotechnics exploded.
> There was no loss of life or injury to anyone, and the property damage outside of the explosives magazine itself was negligible.
> Weather conditions affecting the content of gas shells exploded by the blast may make it desirable for the Army to evacuate temporarily a few civilians from their homes.

This release ran in newspapers and was heard on the radio.

Groves sent a coded message to President Truman, who was with Prime Minister Churchill. Both men were at the Potsdam Conference (July 17–August 2, 1945), waiting to meet with Joseph Stalin to discuss a demand for unconditional surrender from the Japanese. The message read, "Doctor has just returned most enthusiastic and confident that the little boy is as husky as his big brother. The light in his eyes discernible from here to Highhold and I could hear his screams from here to my farm."

Churchill was informed of the Trinity success and congratulated Truman. Churchill had been briefed on the Manhattan Project since it became operational.

On July 24, at Potsdam, Truman quietly told Stalin that the United States "had developed a powerful weapon more potent than anything yet seen in war." Stalin was unimpressed with this news and showed no interest. He was already aware of the atomic bomb project, having learned about it through espionage before Truman became president.

This photograph of the Trinity test was taken by Jack Aeby, using a Perfex 44 camera with 35-mm film. This image provided the basis for the Theoretical Division's earliest calculations of the Trinity weapon's yield. Only one good color photograph was made of the explosion. Autographed and dated on reverse side by Jack Aeby. This is an original print (Los Alamos Photo Laboratory; author's collection).

On the same day, Truman discussed the Trinity explosion with his military advisors and with Churchill. He had concluded that a mainland invasion of Japan would cost too many American lives. He asked his aides to proceed with his wishes to use the atomic bomb.

There is no written letter documenting this historic order. Truman would

Chapter 4. The First Atomic Bomb: Trinity

The much-awaited "Big Three" Postdam Conference was formally opened at 5:00 (Berlin Time) on the afternoon of July 17, 1945. The conference took place in a country estate in the Potsdam area of Berlin. Seen together just before the opening of the conference are Soviet Premier Joseph Stalin, President Harry S. Truman, and Prime Minister Winston Churchill, holding a cigar (U.S. Army Signal Corps, Harry S. Truman Library & Museum).

later write, "The final decision of where and when to use the atomic bomb was up to me. Let there be no mistake about it."

President Truman wrote in his private journal on July 25, 1945:

> We met at 11 A.M. today. That is Stalin, Churchill and the U.S. President. But I had a most important session with Lord

On the back of the "Big Three" photograph, Truman wrote the following: "This is the place I told Stalin about the Atom Bomb, which was exploded July 16, 1945, in New Mexico. He didn't realize what I was talking about!" (Harry S. Truman Library & Museum).

July 25 1945

We met at 11 A.M. today. That is Stalin, Churchill and the U.S. President. But I had a most important session with Lord Mountbatten & General Marshall before that. We have discovered the most terrible bomb in the history of the world. It may be the fire distruction prophesied in the Euphrates Valley Era, after Noah and his fabulous Ark.

Anyway we think we have found the way to cause a disintegration of the atom. An experiment in the New Mexican desert was startling — to put it mildly. Thirteen pounds of the explosive caused the complete disintegration of a steel tower 60 feet high, created a crater 6 feet deep and 1200 feet in diameter, knocked over a steel tower 1/2 mile away and knocked men down 10,000 yards away. The explosion was visible for more than 200 miles and audible for 40 miles and more.

This weapon is to be used against Japan between now and August 10th. I have told the Sec. of War, Mr Stimson, to use it so that military objectives and soldiers and sailors are the target and not women and children. Even if the Japs are savages, ruthless, merciless and fanatic, we as the leader of the world for the common welfare cannot drop this terrible bomb on the old Capitol or the new.

He & I are in accord. The target will be a purely military one and we will issue a warning statement asking the Japs to surrender and save lives. I'm sure they will not do that, but we will have given them the chance. It is certainly a good thing for the world that Hitler's crowd or Stalin's didn't discover this atomic bomb. It seems to be the most terrible thing ever discovered, but it can be made the most useful

HARRY S. TRUMAN LIBRARY
Papers of Harry S. Truman

Chapter 4. The First Atomic Bomb: Trinity

Mountbatten & General Marshall before that. We have discovered the most terrible bomb in the history of the world. It may be the fire destruction prophesied in the Euphrates Valley Era, after Noah and his fabulous Ark. Anyway we think we have found the way to cause a disintegration of the atom. An experiment in the New Mexican desert was startling—to put it mildly. Thirteen pounds of the explosive caused the complete disintegration of a steel tower 60 feet high, created a crater 6 feet deep and 1200 feet in diameter, knocked over a steel tower 7 miles away and knocked men over 10,000 yards away. The explosion was insight for more than 200 miles and audible for 40 miles and more.

This weapon is to be used against Japan between now and August 10th. I have told the Sec. of War. Mr. Stinson to use it so that military objectives and soldiers and sailors are the target and not women and children. Even if the Japs are savages, ruthless, merciless and fanatic, we as the leader of the world for the common welfare cannot drop that terrible bomb on the old capital or the new.

He and I are in accord. The target will be a purely military one and we will issue a warning statement asking the Japs to surrender and save lives. I'm sure they will not do that, but we will have given them the chance. It is certainly a good thing for the world that Hitler's crowd or Stalin's did not discover this atomic bomb. It seems to be the most terrible thing ever discovered, but it can be made the most useful.

The implementation of this "directive" would fall on the shoulders of Colonel Paul Tibbets.

Opposite: **Harry S. Truman diary entry, July 25, 1945. A one-page handwritten entry on his views of using the atomic bomb against Japan (Harry S. Truman Library & Museum).**

Chapter 5

Paul W. Tibbets

Paul Tibbets' father wanted his son to be a medical doctor. Tibbets had come from a family of doctors. However, in university, young Tibbets decided to become a military aviator. His mother supported his decision. In his book, *The Tibbets Story*, Tibbets quotes his mother as saying, "If you want to fly, go right ahead and do it, because I know you will be all right."

Tibbets was an exceptional pilot and quickly rose through the ranks to become a major.

On August 17, 1942, Tibbets, as commander of the 340th Bomb Squadron in the 97th Bombardment Group, along with his crew, flew a B-17 to attack the French city of Rouen. This mission was the first raid made by an American squadron on German-occupied Europe. It dropped 1,100-pound bombs at a locomotive workshop at the Sotteville marshaling yards at Rouen. This was the largest railroad-switching facility in northern France.

Twelve B-17s were on the mission, flying in formation, led by Tibbets. Tibbets and the other 11 B-17s attacked the primary target while another six B-17s attacked a secondary target. The second group was led by a plane called *Yankee Doodle*, with the Eighth Air Force commander, Brigadier General Ira Eaker, on board.

Tibbets also took part in the October 9, 1942, raid on Lille. This was the first time 115 airplanes were in the air on an important bombing mission. Tibbets flew the lead aircraft.

Tibbets implemented tactical bombing strategies in the early war years and was regarded as an outstanding leader, both by the men who served under him and by the higher brass.

In October 1942, Tibbets was involved in Operation Torch. This was the code name for the Allied invasion of Northwest Africa. To accomplish this task, a secret meeting with General Charles Emmanuel Mast, the French commander in Algiers, had to take place in Cherchell, Algeria.

General Carl Spaatz picked his two best pilots: Paul Tibbets and Wayne Connors. The plan was to deliver General Lyman Lemnitzer (head of the Allied forces) and General Mark Clark for that meeting. The meeting was intended to confirm the support of an Anglo-American drive to eradicate the Germans from the southern part of the Mediterranean. Tibbets flew General Clark to Gibraltar. From Gibraltar, a British submarine took General Clark to Cherchell. When

Lieutenant General Ira C. Eaker, commander of the Eighth Air Force. He was the architect of a strategic bombing force that ultimately numbered 40 groups of 60 heavy bombers each, supported by a subordinate fighter command of 1,500 aircrafts. Autographed by General Eaker (U.S. Air Force; author's collection).

the meeting was over, Tibbets and the general flew back. Wayne Connors did the same for General Lemnitzer.

On December 25, 1942, Lieutenant Colonel Paul Tibbets led an 18-airplane attack against Bizerte. The highly successful attack destroyed the port as well as the airfield. News of the raid eventually reached General Jimmy Doolittle.

Left: During World War II, General Lyman L. Lemnitzer was assigned to General Dwight D. Eisenhower's staff, where he helped plan the invasions of North Africa and Sicily. Lemnitzer was one of the senior officers sent to negotiate the Italian Fascist surrender during the secret Operation Sunrise and the German surrender in 1945. Autographed by General Lemnitzer (U.S. Army; author's collection). Right: General Carl Spaatz assumed command of the U.S. Strategic Air Forces in the Pacific as part of the Pacific Theater of Operations, with headquarters on Guam, in July 1945. From this command, Spaatz directed the strategic bombing of Japan, including the atomic bombings of Hiroshima and Nagasaki. Autographed by General Spaatz (U.S. Air Force; author's collection).

General Doolittle went to see Tibbets and told him that operational planning was under way in Algiers. Doolittle wanted the "most experienced bomber man" at his headquarters.

In January 1943, Tibbets, who had now flown 43 combat missions, was assigned as the assistant for bomber operations to Colonel Lauris Norstad, assistant chief of operations of the Twelfth Air Force. Tibbets would later write in his book, "After six months of flying with the best bomber outfit in that part of the world, I was being relegated to a desk to work for one of the most vain and egotistical officers I had ever met."

At one strategic meeting to discuss the resumption of bombing on Bizerte, Tibbets showed his lifelong character. He would always protect his crews regardless of the consequences.

The Germans had rebuilt the harbor installations and repaired the airfield in Bizerte. Colonel Norstad ordered the B-17s to fly at 6,000 feet for their bombing raid. Tibbets told Norstad that, as a bomber expert, flying at 6,000 feet would be disastrous. Norstad felt humiliated in front of the staff officers and said, "It appears that Colonel Tibbets has been flying too much. He may be suffering from battle fatigue."

Chapter 5. Paul W. Tibbets

Left: General Mark W. Clark made a covert visit to French North Africa to meet with pro–Allied officers of the Vichy French forces. His pilot was Paul Tibbets. Autographed by General Clark (U.S. Army; author's collection). *Right:* General Lauris Norstad, deputy chief of air staff at Army Air Force Headquarters. Autographed by General Norstad (U.S. Air Force; author's collection).

Tibbets answered Norstad, "I'll tell you what I'm prepared to do, Colonel. I'll lead that raid myself at 6,000 feet if you will come along as my co-pilot."

Norstad, who had never flown in combat missions, backed down. He was not prepared to fly over a heavily defended enemy target at any altitude.

Tibbets applied his past experience in his recommendations for the attack on Bizerte. The Germans were using 88-millimeter anti-aircraft guns and had defense installations all over Bizerte. Tibbets said that the best altitude for bombing Bizerte was around 14,000–20,000 feet.

The following day, the B-17s flew to Bizerte and completed their bombing raid at 20,000 feet. The results were excellent, with no losses.

Afterward, General Hap Arnold needed an experienced bombardment officer to help develop the new Boeing B-29s. So Tibbets had a new job.

In February 1943, the Army transferred Tibbets back to the United States to supervise the testing and help train flight instructors for the newly built B-29 "Superfortresses." These were the largest and best-equipped bombers in the world.

Chapter 6

Preparations for the Atomic Bombings

In his memoir *Now It Can Be Told*, General Leslie Groves said that Tibbets had been selected to train the crews because "he was a superb pilot of heavy planes, with years of military flying experience, and was probably as familiar with the B-29 as anyone in the service."

On September 1, 1944, Lieutenant Colonel Paul Tibbets was called to a special meeting at the U.S. Army Second Air Force Headquarters in Colorado Springs. At that meeting, he was introduced to General Uzal Ent, commander of the Second Air Force; Colonel John Lansdale, chief of security for the Manhattan Project; Navy Captain William "Deak" Parsons, associate director of the Los Alamos Laboratory; and physicist Norman Ramsey. Tibbets was told to organize a combat unit whose purpose was to deliver a new type of explosive to an unknown target. This unit would be self-sufficient and would be able to operate anywhere in the world.

Dr. Norman Ramsey, a Columbia University professor, gave Tibbets a brief explanation of the principles of atomic fission and the developing bomb. He was told that when the bomb was finally assembled, it would have the explosive power of thousands of tons of TNT. Ramsey told Tibbets that he believed the bomb would shorten the war.

Dr. Ramsey explained that over 100,000 people had been working on this project in secrecy in several laboratories in the United States. Its name was the Manhattan Project.

Captain Parsons told Tibbets that the bomb being built would weigh 9,000 pounds and be 12 feet long with a diameter of 28 inches. Parsons further emphasized that the aircraft carrying the bomb would have to be at least eight miles away from the bomb blast; otherwise, the shock waves would cause severe structural damage to the aircraft.

At the meeting, Colonel Lansdale told Tibbets, "Colonel, I want you to understand one thing. Security is first, last, and always. You will commit as little as possible to paper. You will tell only those who need to know what they must know to do their jobs properly."

General Ent assigned Tibbets to be commander of the 393rd Heavy Bombardment Squadron, based in Harvard, Nebraska.

Tibbets' new group would be called the 509th Composite Group. It would be

Chapter 6. Preparations for the Atomic Bombings 51

a completely self-contained outfit with approximately 1,800 men and 15 B-29s. The personnel he would need to select would include ordnance, engineering, maintenance, military police, medical units, and specialists in radioactivity. Tibbets would be responsible for finding a suitable training location, choosing his team and locating all needed equipment. All this with complete secrecy!

Tibbets would have the authority to requisition anything he needed to complete his assignment under the code name "Silverplate."

Tibbets was also offered his choice from three airbases for a center of operation. They were Wendover in Utah, Mountain Home in Idaho, and Great Bend in Kansas.

Tibbets flew first to Wendover and decided it was an ideal location. The isolated town of Wendover had a population of about 100 people. The nearest city was Salt Lake City, which was 125 miles away. Tibbets never visited the two other locations.

Headquarters were set up on September 8, 1944, with 15 B-29s and flight and ground crews.

Security was the highest priority. On the base there were signs reminding everyone about tight security wherever they went. Beside the exit gate, a large sign (see next page) read:

Wendover, Utah, headquarters of the 509th Composite Group. Photograph autographed by Charles Albury, who participated in both atomic bomb missions (U.S. Air Force; author's collection).

Left to right: Major Thomas W. Ferebee, bombardier; Colonel Paul W. Tibbets, pilot; Captain Theodore J. Van Kirk, navigator; and Captain Robert Lewis, co-pilot. Officer crew of the *Enola Gay*, the aircraft that made the historic flight over Hiroshima to drop the first atomic bomb. Autographed by Theodore Van Kirk (U.S. Air Force; author's collection).

> WHAT YOU HEAR HERE
> WHAT YOU SEE HERE
> WHEN YOU LEAVE HERE
> LET IT STAY HERE

Special agents were employed to bug telephones, open all mail, and eavesdrop on conversations. Anyone found to have disregarded Tibbets' code of silence was immediately shipped out to Alaska.

A majority of the personnel came from the 393rd Bomb Squadron. These men and women would form the core of the new unit. Lieutenant Colonel Thomas Classen had been their commander. Tibbets decided to let Classen become deputy commander of the 509th. He would be responsible for training the crews with the help of Tibbets' crew. Tibbets would make all the major policy decisions.

Tibbets gathered personnel whom he had worked with in the past—men he trusted, who had superior skills and were battle tested. These included Tom Ferebee, his best bombardier, with 63 combat missions; George Caron, tail gunner; Theodore Van Kirk, navigator; Bob Lewis, co-pilot; and Wyatt Duzenbury,

Chapter 6. Preparations for the Atomic Bombings

Tom Ferebee, bombardier; Paul Tibbets, pilot; Theodore Van Kirk, navigator on *Enola Gay*. Autographed by Tom Ferebee (U.S. Army; author's collection).

flight engineer. This core group was assigned to Tibbets' own plane and would be responsible for training their counterparts in the other aircraft.

To supervise security, Tibbets brought in Lieutenant Colonel Hazen Payette. In addition, Major William L. "Bud" Uanna came with 30 agents. These agents would be the police force at Wendover. They arrived with detailed files on every member of the 393rd Bomb Squadron and every man whom Tibbets had personally selected from air bases in the United States and Europe. These files represent the most thorough secret investigation of individuals that had ever been performed up to that time.

Tibbets also brought in Kermit Beahan, bombardier; James Van Pelt, navigator; Jacob Beser, radar countermeasures officer; and pilots Charles Sweeney and Don Albury.

All 15 B-29 crews were trained at Wendover Air Base. The crews were also making runs at a bombing range on Sandy Beach in the Salton Sea Test Range, California. There a manmade lake was constructed and used as target practice. Over 200 conventional practice bombs were used for the training by the bombardiers to get them acquainted with the Norden bombsight at higher altitudes.

Beser, one of the highest-rated radar officers in the Army, was sent to Los Alamos for classified training. He was taught how to monitor enemy radar to see whether it was trying to jam or detonate the mechanism of the bomb. Beser

Lieutenant Colonel Hazen J. Payette, group intelligence officer, briefing Major Thomas W. Ferebee, Colonel Paul W. Tibbets, and Captain Theodore J. Van Kirk before the Hiroshima atomic bomb mission. Autographed by Paul Tibbets (U.S. Air Force; author's collection).

also had to learn how the bomb's inner firing mechanism (including its built-in mini-radar system) worked. On every flight, Beser would bring on board 300 pounds of equipment consisting of his spectrum analyzers, direction finder, search receivers and antennas.

Over the next few months, Tibbets would make three visits to Robert Oppenheimer. Tibbets wanted to know exactly what to expect when the bomb was detonated. He especially asked Oppenheimer what effect the shock waves would have on the plane.

Oppenheimer told Tibbets, "Turn 155 degrees as fast as you can and you'll be able to put yourself the greatest distance from where the bomb exploded." Flying straight would cause the wings and the underbelly of the aircraft to be exposed to tremendous shock waves. At a 155 degree angle to the detonation, the aircraft would have a much smaller vulnerable area, and would survive.

Tibbets was originally told that the B-29 must be at least eight miles away from the bomb blast. The bomb blast would theoretically occur 43 seconds after the bomb was dropped from an altitude of about 31,000 feet, when it reached 1,890 feet above ground level. Shock waves travel at the speed of sound (1,125

Chapter 6. Preparations for the Atomic Bombings

feet per second), which meant the first shock wave would reach the *Enola Gay* about 60 seconds after detonation. The time required to be eight miles away is two minutes. Tibbets would have only 40–42 seconds to make a 155-degree turn and escape the blast radius.

Tibbets had 1,800 men and 15 B-29s under his direct command. All crews would practice with dummy bombs having the same shape, size, and weight as the real bombs. They trained at precision bombing, flying at different altitudes, and flying over water at night. Tibbets also ordered every pilot to practice making a sharp 155-degree turn away from predetermined aiming points at an altitude of 30,000 feet.

One of the major advantages of Wendover (code name "W-47") as a base for the 509th Composite Group, and the drop-test programs, was that the base was situated near a main railroad line. A spur ran directly into the base. Here, sealed containers of classified bomb parts, known as "Pumpkins" (large inert copies of "Little Boy" and "Fat Man"), could be delivered to the base without security risk.

Based on considerable combat bombing experience, along with what Captain William Parsons had told him, Tibbets knew that in any attack on a Japanese target, the B-29s would have to be cruising at 30,000 feet or higher. He knew this altitude would prevent most anti-aircraft fire.

To accomplish this task, Tibbets ordered that all B-29s be completely stripped of armaments and armor plating. This saved 7,000 pounds of installed weight and increased speed to escape the shock waves. Only the tail gunner would be in a position to shoot down any incoming enemy aircraft.

Seventeen production Silverplate aircraft were ordered in August 1944 to allow the 509th Composite Group to train with the type of aircraft they would have to fly in combat and for the 216th Army Air Forces Base Unit to test bomb configurations. These were followed by 28 more aircraft that were ordered in February 1945 for operational use by the 509th Composite Group. These latest models came with fuel-injected engines and electronically controlled reversible propellers, as well as other improved features.

Tibbets requested that the new B-29s be ordered from the Martin plant in Omaha, with very little armor plate and with only tail guns for armament. Critically important were the new bomb bay doors. These were designed to automatically close in two seconds after the release of a bomb. (This would allow the plane to make the 155-degree turn even faster.)

Tibbets was promoted from lieutenant colonel to full colonel in January 1945.

On December 28, 1944, Groves sent for Tibbets. Tibbets was told that the 509th Composite Group should be ready on June 15 to deliver an atomic strike. Groves gave the government's opinion on how the targets should be selected:

> The targets chosen should be places the bombing of which would most adversely affect the will of the Japanese people to continue the war. Beyond that, they should be military in nature, consisting either of important headquarters or troop concentrations, or centers of production of military equipment and supplies. To enable us to assess accurately the effects of the bomb, the targets should not have been previously damaged by air attacks. It is also desirable that the first target be of such size that the damage would be confined within it, so that we could more definitely determine the power of the bomb.

On December 30, 1944, Grove sent a note to General George C. Marshall, Army Chief of Staff:

> It is now reasonably certain that our operations plans should be based on the gun-type bomb, which, it is estimated, will produce the equivalent of a ten thousand ton TNT explosive. The first bomb, without previous full scale test, which we do not believe will be necessary should be ready about 1 August, 1945.

On January 6, 1945, Tibbets sent 10 complete crews to Cuba. They continued their training at Batista Field, 12 miles from Havana. Here a training task force had been established for extensive long-range flying over water.

In the beginning of March 1945, General Marshall asked Groves to pick the targets.

On March 9, 1945, General Curtis LeMay ordered a fire-bombing attack on Tokyo. In the final briefing, he told the crews, "You're going to deliver the biggest firecracker the Japanese have ever seen!"

At midnight, 325 B-29s took-off from Guam for "Operation Meetinghouse." They flew at low altitude, stripped of any armaments. Each plane had bomb bays filled with incendiary bombs (M-69 napalm bombs). The raid started at 12:00 a.m. By 3:30 a.m., it was over. Approximately 2,000 tons of these bombs were dropped on Tokyo. Fourteen aircraft with crews were lost.

Tokyo's air defenses were caught off-guard. One hundred thousand people died; about half a million were injured. Two hundred and fifty thousand buildings were destroyed in a 16-square-mile area. President Truman would later remark, "Despite their heavy losses at Okinawa and the firebombing of Tokyo, the Japanese refused to surrender. The saturation bombing of Japan took much fiercer tolls and wrought far and away more havoc than the atomic bomb. Far and away. The firebombing of Tokyo was one of the most terrible things that ever happened, and they didn't surrender after that although Tokyo was almost completely destroyed."

In late March 1945, Second Lieutenant Morris R. Jeppson of the First Ordnance Squadron was brought to Los Alamos to learn everything about the new secret proximity-fuse firing mechanism. The crew of Lewis, Ferebee and Van Kirk continued to do practice bombing runs.

At 3:35 p.m., April 12, 1945, Franklin Delano Roosevelt was pronounced dead. At 5:45 p.m., Harry S. Truman was sworn in by Justice Harlan Stone. At this time, he was still unaware of the existence of the Manhattan Project.

In his initial address before Congress, on April 16, 1945, Truman said, "Tokyo rocks under the weight of our bombs.... I want the entire world to know that this direction must and will remain—unchanged and unhampered. Our demand has been and it remains—unconditional surrender."

On April 24, 1945, Secretary of War Henry L. Stimson wrote a letter to President Truman, urging that they meet.

President Truman met Stimson on April 25. At the request of General Stimson, General Groves was invited to the meeting. Stimson informed Truman about the Manhattan Project and explained the basic theory behind the making of an atomic bomb.

Chapter 6. Preparations for the Atomic Bombings

White House Central Files
Confidential Files

SECRET

WAR DEPARTMENT
WASHINGTON

April 24, 1945.

Dear Mr. President:

I think it is very important that I should have a talk with you as soon as possible on a highly secret matter.

I mentioned it to you shortly after you took office but have not urged it since on account of the pressue you have been under. It, however, has such a bearing on our present foreign relations and has such an important effect upon all my thinking in this field that I think you ought to know about it without much further delay.

Faithfully yours,

Henry L. Stimson
Secretary of War.

The President,
The White House.

DECLASSIFIED
E. O. 11652, Sec. 3(E) and 5(D) or (E)
OSD letter, April 12, 1974
By NLT-___, NARS Date 2-2-76

SECRET

Harry S. Truman was sworn in as the new president on April 12, 1945. Nearly two weeks later, Henry Stimson, secretary of war, sent a top-secret letter to President Truman on April 24, 1945. This document is initialed by Truman "HST" and notes "Put on list tomorrow, Wed. 25." On April 25, Stimson (for the first time) told the president about the Manhattan Project. This document was declassified on April 12, 1974 (Harry S. Truman Library & Museum).

A recently declassified memorandum shows exactly what Stimson told President Truman:

> Within four months we shall in all probability have completed the most terrible weapon ever known in human history, one bomb of which could destroy a whole city.
> Although we have shared its development with the United Kingdom, physically the U.S. is at present in the position of controlling the resources with which to construct and use it and no other nation could reach this position for some years. Nevertheless, it is practically certain that we could not remain in this position indefinitely.

Stimson told Truman that using the bomb against Japan would probably shorten the war. General Groves agreed.

At the end of the meeting, Truman agreed to form a special panel, to be known as the Interim Committee. This committee advised Truman on all aspects of atomic energy.

On May 8, 1945, at 9:00 a.m., President Harry S. Truman gave a live broadcast. He announced "Victory in Europe" Day, following the surrender of Germany on May 7, 1945.

On May 9, 1945, Tibbets flew to the Martin bomber plant in Omaha, Nebraska. He walked up and down the assembly line and picked out a new B-29 that he would personally use for the first atomic strike.

On May 25, 1945, the Joint Chiefs of Staff began the formal planning for an attack on Japan. The proposed assault would involve at least two million American soldiers. The plan (code named "Operation Downfall") was to force the Japan's unconditional surrender. Operation Downfall was further divided into two campaigns. The first attack on Kyushu Island would have occurred in November 1945, under the code name "Olympic," and would have involved 815,548 troops. The second invasion would have occurred in April 1946 at Honshu Island, near Tokyo, under the code name "Coronet." This operation would have used 171,646 troops. (These are the actual estimates that Truman approved.)

Had Operation Olympic been necessary, it would have been the largest amphibious assault in history. Its goal was to secure advanced airfields and ports for an invasion into the industrial heart of Japan.

Truman was told by the Joint Chiefs that an invasion of mainland Japan could take a year to complete and result in 250,000 to 500,000 U.S. casualties.

Meanwhile, Groves and the Target Committee had decided that the bombing targets (in order of preference) should be Kyoto, Hiroshima, Yokohama, and Kokura. All were war industry and military centers.

Secretary of War Stimson opposed Kyoto as a primary target because "it was an historical city and one that is of great religious significance to the Japanese." Kyoto was later dropped because of its historical significance, as was Kokura. In their place, Niigata and Nagasaki were added to the list.

The final Interim Committee meeting took place in the Pentagon on May 28, 1945. At this secret meeting, ordered by Stimson, final plans for the atomic bombings were discussed in great detail. In attendance were Groves, Tibbets, and Beser, as well as scientists Oppenheimer, Fermi, Lawrence (in addition to Arthur Compton). All were given large-scale maps and reconnaissance photographs of

Chapter 6. Preparations for the Atomic Bombings

possible targets. (Due to the fact that everything was secret, no classified document exists as to what was discussed.)

On June 11, 1945, seven scientists from the Chicago Laboratory submitted a secret document, which would be known as the Franck Report. It was signed by James Franck, Donald J. Hughes, James J. Nickson. Eugene Rabinowitch, Glenn T. Seaborg, J.C. Steans, and Leo Szilard. Its title:

> Report of the Committee on Political and Social Problems
> Manhattan Project "Metallurgical Laboratory"
> University of Chicago, June 11, 1945
> (The Franck Report)

The report basically petitioned the secretary of war not to use an atomic bomb on any Japanese city. Instead, the authors recommended "demonstrat[ing] before observers from many countries in an uninhabited area."

From the summary of this report:

> Nuclear bombs cannot possibly remain a "secret weapon" at the exclusive disposal of this country, for more than a few years. The scientific facts on which their construction is based are well known to scientists of other countries. Unless an effective international control of nuclear explosives is instituted, a race of nuclear armaments is certain to ensue following the first revelation of our possession of nuclear weapons to the world. Within ten years other countries may have nuclear bombs, each of which, weighing less than a ton, could destroy an urban area of more than ten square miles. In the war to which such an armaments race is likely to lead, the United States, with its agglomeration of population and industry in comparatively few metropolitan districts, will be at a disadvantage compared to the nations whose population and industry are scattered over large areas.
>
> We believe that these considerations make the use of nuclear bombs for an early, unannounced attack against Japan inadvisable. If the United States would be the first to release this new means of indiscriminate destruction upon mankind, she would sacrifice public support throughout the world, precipitate the race of armaments, and prejudice the possibility of reaching an international agreement on the future control of such weapons.
>
> Much more favourable conditions for the eventual achievement of such an agreement could be created if nuclear bombs were first revealed to the world by a demonstration in an appropriately selected uninhabited area.

Manhattan Project authorities were so outraged by this report that some sentences in all copies—including the originals in the National Archives—were permanently censored with ink.

Four days later, the Interim Committee met and discussed this report at length. It concluded, "We can propose no technical demonstration likely to bring an end to the war; we see no acceptable alternative to direct military use."

Truman's decision to use the atomic bomb was based on his desire to finally end the war and substantially reduce casualties on both sides. In part, Truman's observations of the Battle of Okinawa convinced him that a mainland invasion launched by American and Allied troops would result in horrific casualty figures.

In the Battle of Okinawa (code named "Operation Iceberg"), the U.S. 10th Army had 182,821 personnel under its command, combined with the 1st, 2nd, and 6th Divisions of the Marine Corps, to fight on the island. The Allies were planning

A REPORT TO THE SECRETARY OF WAR, JUNE 1945

   I. Preamble

The only reason to treat nuclear power differently from all other
developments in the field of physics is the possibility of its use
as a means of political pressure in peace and sudden destruction in
war. All present plans for the organization of research, scientific
and industrial development, and publication in the field of nucleonics
are conditioned by the political and military climate in which one
expects those plans to be carried out. Therefore, in making suggestions
for the postwar organization of nucleonics, a discussion of political
problems cannot be avoided. The scientists on this Project do not
presume to speak authoritatively on problems of national and
international policy. However, we found ourselves, by the force of
events, during the last five years, in the position of a small group of
citizens cognizant of a grave danger for the safety of this country as
well as for the future of all the other nations, of which the rest of
mankind is unaware. We therefore feel it is our duty to urge that the
political problems, arising from the mastering of nuclear power, be
recognized in all their gravity, and that appropriate steps be taken for
their study and the preparation of necessary decisions. We hope that
the creation of the Committee by the Secretary of War to deal with all
aspects of nucleonics, indicates that these implications have been
recognized by the government. We believe that our acquaintance with the
scientific elements of the situation and prolonged preoccupation with its
world-wide political implications, imposes on us the obligation to offer
to the Committee some suggestions as to the possible solution of
these grave problems.

   Scientists have often before been accused of providing new weapons for
the mutual destruction of nations, instead of improving their well-being.
It is undoubtedly true that the discovery of flying, for example, has
so far brought much more misery than enjoyment and profit to humanity.
However, in the past, scientists could disclaim direct responsibility
for the use to which mankind had put their disinterested discoveries. We
feel compelled to take a more active stand now because the success which
we have achieved in the development of nuclear power is fraught with
infinitely greater dangers than were all the inventions of the past. All
of us, familiar with the present state of nucleonics, live with the
vision before our eyes of sudden destruction visited on our own country,
of a Pearl Harbor disaster repeated in thousand-fold magnification in
every one of our major cities.
   In the past, science has often been able to provide also new methods
of protection against new weapons of aggression it made possible, but
it cannot promise such efficient protection against the destructive use
of nuclear power. This protection can come only from the political
organization of the world. Among all the arguments calling for an
efficient international organization for peace, the existence of
nuclear weapons is the most compelling one. In the absence of an
international authority which would make all resort to force, in

In June 1945, a group of scientists at the University of Chicago prepared a secret report (the "Franck Report") arguing against the use of the atomic bomb. The group was headed by James Franck and included Donald J. Hughes, J. J. Nickson, Eugene Rabinowitch, Glenn T. Seaborg, J. C. Stearns and Leo Szilard. Page 1 of 8. Autographed by Glenn Seaborg (author's collection).

to use Kadena Air Base on the large island of Okinawa as a base for Operation Downfall.

The amphibious assault on Okinawa was the largest in the Pacific Theater of World War II. It started on April 1, 1945, and ended 82 days later. The ferocity of the fighting, the intensity of Japanese kamikaze attacks, and the sheer number of

Chapter 6. Preparations for the Atomic Bombings

Portrait of Niels Bohr, James Franck, Albert Einstein, and I. I. Rabi. Autographed by all (Smithsonian Libraries, Washington, D.C.).

Allied ships and armored vehicles that assaulted the island made this battle one of the bloodiest in the Pacific.

The U.S. National Archives records the deaths as follows: 14,009 American soldiers, 149,425 Okinawan civilians, 77,166 Japanese soldiers, 4,907 naval personnel, and 2,938 Marine Corps personnel. Several thousands who later died from their wounds were not counted. The huge casualty rate for the Okinawan civilians was due to the Japanese tactic of using the Okinawans as human shields.

At sea, 368 Allied ships (including 120 amphibious craft) were damaged, while another 36 (including 15 amphibious ships and 12 destroyers) were sunk. In addition, the United States lost 768 planes, while the Japanese lost over 1,100 planes.

Through intelligence reports, the United States knew that Japan was in the process of bringing 2,000,000 troops from China back to defend the mainland.

The expectation of the American people after the attack on Pearl Harbor, and after hearing the survivors' accounts of the Bataan Death March, was to see an unconditional Japanese surrender. Only a major calamity would stop an American invasion of imperial Japan.

Chapter 7

Tinian

Tinian was captured on August 1, 1944. Tinian was about 40 square miles in size and located approximately 1,500 miles due south of Tokyo. A round-trip mission from Tinian to Tokyo would take 12 hours. This proximity to Japan was the main reason why Tibbets chose Tinian to be the headquarters of the 509th Composite Group.

On May 6, 1945, 800 ground crew members of the 509th Composite Group left Wendover for embarkation at Seattle. This transition was followed by the bomber crews flying to the Pacific.

In June 1945, the 509th Composite Group relocated to Tinian (code name

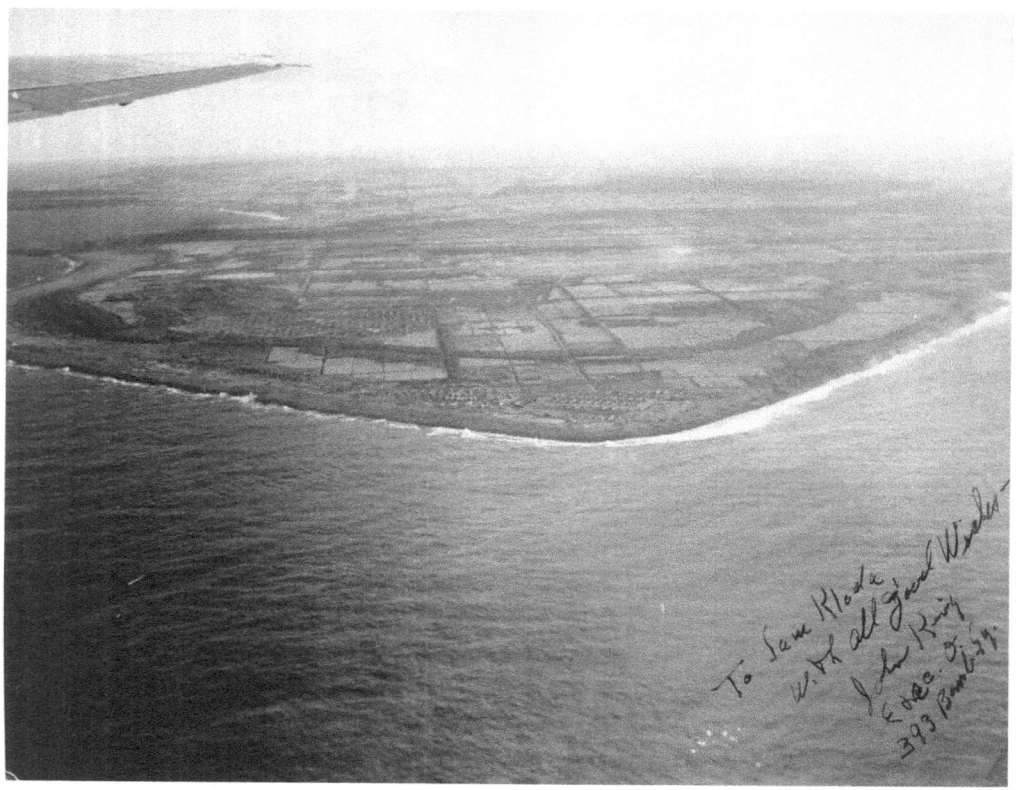

Air approach to Tinian. Autographed by Captain John A. King, executive officer of the 393rd Bombardment Squadron (U.S. Air Force; author's collection).

Assembly building of the 509th Composite Group on Tinian, Mariana Islands. This is where the bombs were assembled. Autographed by John A. King, executive officer of the 393rd Bombardment Squadron (U.S. Air Force; author's collection).

First Ordnance Squadron area of the 509th Composite Group on Tinian, Mariana Islands. Autographed by John A. King, executive officer of the 393rd Bombardment Squadron (U.S. Air Force; author's collection).

"Destination O"), Mariana Islands, in the Pacific Ocean for final preparation for its bombing missions. Soon after the island's seizure by the United States, construction began on the largest airbase of World War II, which covered almost the entire island. The base had 40,000 personnel and hundreds of aircraft.

Tibbets was assigned the North Field of Tinian, which had four 8,500-foot runways.

North Field had a total of 269 B-29s.

Phillip Morrison was a physicist who worked in the critical assemblies group of the Weapon Physics Division and participated in dangerous critical experiments. After the Trinity test, Morrison traveled to Tinian as a pit team leader for Project Alberta, where he helped oversee the assembly of the "Fat Man" plutonium bomb.

In Richard Rhodes' book, *The Making of the Atomic Bomb*, Morrison gives a vivid description of Tinian:

> Tinian is a miracle. Here, 6,000 miles from San Francisco, the United States armed forces have built the largest airport in the world. A great coral ridge was half-levelled to fill a rough plain, and to build six runways, each an excellent 10-lane highway, each almost two

Photo of Tinian. Autographed by James N. Price, airplane commander of *Some Punkins* (U.S. Air Force; author's collection).

miles long. Beside these runways stood in long rows, the great silvery airplanes. They were not by the dozen, but by the hundred. From the air this island, smaller than Manhattan, looked like a giant aircraft carrier, its deck loaded with bombers.

The 509th Composite Group now contained the 393rd Bomber Squadron, the 320th Troop Carrier Squadron, the 390th Air Service Group, the 603rd Air Engineering Squadron, the 1027th Air Materiel Squadron, and the First Ordnance Squadron. The First Ordnance Squadron would have the job of handling the atomic bombs when they arrived and then be responsible for their entry into the bomb bays.

The *Indianapolis* left Hunter's Point, the naval shipyard in San Francisco, for Hawaii. As of July 16, it was the fastest cruiser available for a dangerous mission. Only a few days before, Captain Charles Butler McVay III had received his orders from Rear Admiral William Purnell and Captain Parsons. Parsons' instructions to McVay were as follows:

> You will sail at high speed to Tinian where your cargo will be taken off by others. You will not be told what the cargo is, but it is to be guarded even after the life of your vessel. If she goes down, save the cargo at all costs, in a lifeboat if necessary. Every day you save on your voyage will cut the length of the war by just that much.

The *Indianapolis*, with its 1,200-man crew, headed toward Pearl Harbor and then Tinian.

On July 17, 1945, the 509th and Project Alberta commenced the program of Special Bomb Operations. The program included the following major activities:

1. Practice drops of Pumpkins on targets in the Japanese Empire;
2. Assembly, flight, and drop tests of Little Boy bombs in the Mariana Islands area;
3. Assembly, flight and drop tests of Fat Man bombs in the Mariana Islands area; and
4. Combat drops of the Special Bombs of Fat Man on targets in the Japanese Empire.

On July 19, Tibbets selected ten crews to fly the first bombing missions over Japan. Each crew flew a preselected route, against a different aiming point. This approach was meant to give the crews more combat experience; more important, the Japanese would become accustomed to seeing a lone B-29 flying overhead and then dropping a single bomb.

Colonel Tibbets was specifically forbidden to fly any of these missions over Japan.

Thirteen raids were conducted, covered by written orders: Special Bombing Mission No. 1 to No. 12, as well as Special Bombing Mission No. 14.

The First Ordnance Squadron was responsible for loading the Pumpkins into the B-29s.

For bombing missions over Japan, each airplane would fly alone to its target city, drop its bomb and return to Tinian. Total flight time for the round trip would require 12–14 hours. Seventeen cities were selected by General LeMay for these high-altitude precision bombing practices under realistic combat conditions.

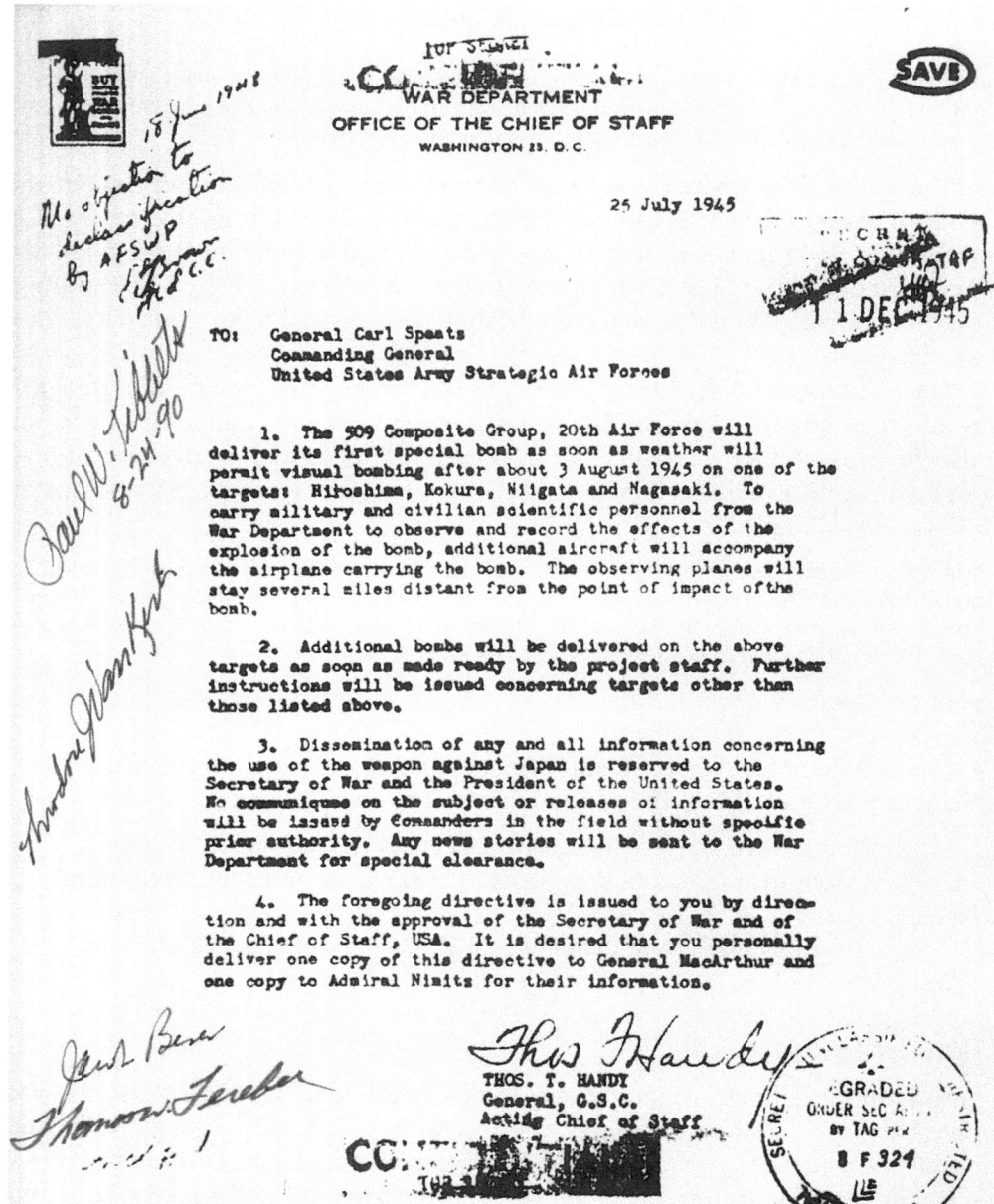

Top-secret letter from General Thomas T. Handy to General Carl Spaatz. At this time, the United States was preparing to drop atomic bombs on Japan. General Spaatz, the commander of the United States Strategic Air Forces in the Pacific, said he wanted his orders in writing. These orders directed him to drop the atomic bomb. Autographed by Paul Tibbets, Theodore Van Kirk, Jacob Beser, and Thomas Ferebee (author's collection).

On July 23, General Groves briefed General Spaatz on the atomic bomb. Now commander of the Strategic Air Force, it would be his duty to carry out the orders of the Target Committee. At the meeting was General Thomas T. Hardy, acting chief of staff while General Marshall was in Potsdam. Spaatz told Hardy, "If I'm going to kill 100,000 people I'm not going to do it on verbal orders. I want a piece of paper."

The document was drafted on July 23 by Groves and presented to Spaatz on July 29.

Admiral Chester Nimitz had been aware of this plan for some time, but General Douglas MacArthur has been deliberately kept in the dark up to this point.

As instructed, General Spaatz visited General MacArthur and handed him a copy of the letter. MacArthur disagreed with the atomic bombing of Japan. He wanted to be commander of a Normandy–type amphibious landing against Japan. MacArthur believed that the potential casualties in this scenario would run into the hundreds of thousands, but he told Spaatz that losses would not be more than 100,000.

General MacArthur now knew that Operation Downfall will not proceed. He would not see his dream of conquering Japan by a land and sea invasion. In addition, he had been formally advised about the existence of the new weapon.

Chapter 8

The Potsdam Declaration

Held near Berlin, the Potsdam Conference (July 17–August 2, 1945) was the last of the World War II meetings held by the "Big Three" heads of state. Featuring American President Harry S. Truman, British Prime Minister Winston Churchill (and his successor, Clement Attlee) and Soviet Premier Joseph Stalin, the talks established a Council of Foreign Ministers and a central Allied Control Council for administration of Germany. The leaders arrived at various agreements on the German economy, punishment for war criminals, land boundaries and reparations.

Postwar Europe was the primary concern of United States, Great Britain and China. On July 26, 1945, they issued a declaration demanding "unconditional surrender" from Japan.

This is the "Proclamation Defining Terms for Japanese Surrender," issued at Potsdam on July 26, 1945:

> 1. We—the President of the United States, the President of the National Government of the Republic of China, and the Prime Minister of Great Britain, representing the hundreds of millions of our countrymen, have conferred and agree that Japan shall be given an opportunity to end this war.
>
> 2. The prodigious land, sea and air forces of the United States, the British Empire and of China, many times reinforced by their armies and air fleets from the west, are poised to strike the final blows upon Japan. This military power is sustained and inspired by the determination of all the Allied Nations to prosecute the war against Japan until she ceases to resist.
>
> 3. The result of the futile and senseless German resistance to the might of the aroused free peoples of the world stands forth in awful clarity as an example to the people of Japan. The might that now converges on Japan is immeasurably greater than that which, when applied to the resisting Nazis, necessarily laid waste to the lands, the industry and the method of life of the whole German people. The full application of our military power, backed by our resolve, will mean the inevitable and complete destruction of the Japanese armed forces and just as inevitably the utter devastation of the Japanese homeland.
>
> 4. The time has come for Japan to decide whether she will continue to be controlled by those self-willed militaristic advisers whose unintelligent calculations have brought the Empire of Japan to the threshold of annihilation, or whether she will follow the path of reason.

Chapter 8. The Potsdam Declaration

5. Following are our terms. We will not deviate from them. There are no alternatives. We shall brook no delay.

6. There must be eliminated for all time the authority and influence of those who have deceived and misled the people of Japan into embarking on world conquest, for we insist that a new order of peace, security and justice will be impossible until irresponsible militarism is driven from the world.

7. Until such a new order is established and until there is convincing proof that Japan's war-making power is destroyed, points in Japanese territory to be designated by the Allies shall be occupied to secure the achievement of the basic objectives we are here setting forth.

8. The terms of the Cairo Declaration shall be carried out and Japanese sovereignty shall be limited to the islands of Honshu, Hokkaido, Kyushu, Shikoku and such minor islands as we determine.

9. The Japanese military forces, after being completely disarmed, shall be permitted to return to their homes with the opportunity to lead peaceful and productive lives.

10. We do not intend that the Japanese shall be enslaved as a race or destroyed as a nation, but stern justice shall be meted out to all war criminals, including those who have visited cruelties upon our prisoners. The Japanese Government shall remove all obstacles to the revival and strengthening of democratic tendencies among the Japanese people. Freedom of speech, of religion, and of thought, as well as respect for the fundamental human rights shall be established.

11. Japan shall be permitted to maintain such industries as will sustain her economy and permit the exaction of just reparations in kind, but not those which would enable her to re-arm for war. To this end, access to, as distinguished from control of, raw materials shall be permitted. Eventual Japanese participation in world trade relations shall be permitted.

On the same date that the Potsdam Declaration was issued, the USS *Indianapolis* delivered "Little Boy" to Tinian Island. With the bomb came the firing mechanism, a cylinder made of lead to prevent the escape of radiation, and a slug of uranium (U-235). A second slug of U-235 arrived via Hamilton Air Force Base in California. Two other B-29s, with plutonium (Pu-239) for the second bomb, also arrived.

Thomas Ferrell, deputy director of the Manhattan Project, was told to sign a receipt showing the transfer of responsibility for the components of the "Little Boy." Three copies were signed. The U.S. government kept two copies, which were eventually lost! Ferrell kept his copy in his wallet until his death. Fortunately, this valuable artifact was donated to the United States Army Heritage and Education Center.

Unfortunately, four days after its arrival at Tinian, on July 30, 1945, the USS *Indianapolis* (then en route to the Philippines) was torpedoed by the Japanese submarine I-58, commanded by Lieutenant Commander Mochitsura Hashimoto. The I-58 fired six torpedoes, each carrying 1,210 pounds of explosives—enough to

There was an official receipt signed by Thomas Ferrell on receiving the first atomic bomb (United States Army Heritage and Education Center, Carlisle, Pennsylvania).

take out a city block. Two torpedoes struck the *Indianapolis*, and it sank 12 minutes later.

In August 2015, Curt Smith, a reporter for the *Lansing State Journal*, interviewed Dick Thelen. Thelen, a gunner's mate, was a survivor from the USS *Indianapolis*. At the time of its sinking, the cruiser had a crew of 1,197. Only 317 survived.

Chapter 8. The Potsdam Declaration

USS *Indianapolis* (CA-35). Pre-war photo, date unknown. Autographed by USS *Indianapolis* survivor Dick Thelen (NavSource-Photographic History of the U. S. NAVY).

Thelen recalled the events: "About 900 men ended up in the ocean. The sharks began to pick off sailors one by one. Thirsty and floating about in 100-degree heat, many gave in to drinking salt water. You gulp salt water on an empty stomach and your eyeballs pop out, foam comes to your mouth and within hours you're dead."

On August 3, 1945, ships arrived to pick up survivors. They were all suffering from severe leg cramps, dehydration, exhaustion, and raw skin rashes caused by being in salt water for five days. On hearing this disturbing news, the Army Air Forces ordered almost 1,000 bombers to attack 12 Japanese cities.

Besides air power, the United States relied on the Office of War Information (OWI). OWI was tasked with using information warfare to promote distrust of Japanese military leaders, encourage surrender and disseminate information by radio and leaflets.

The Japanese were constantly using Radio Tokyo for their propaganda messages to Allied airmen, soldiers and marines. Their programs were broadcast in the South Pacific and North America to demoralize Allied forces abroad and their families at home by emphasizing troops' wartime difficulties and military losses. The OWI used psychological warfare by bombarding Japan with similar radio messages through its 50,000-watt standard-wave station on Saipan, Radio KSAI.

On the day after the Allied call for "unconditional surrender" by the Japanese, the OWI produced three million notices (leaflets written in Japanese). These leaflets warned civilians to evacuate 35 Japanese cities that were scheduled to be bombed in the next few days.

During many of the bombing missions, propaganda leaflets printed by the U.S. Army were dropped by B-29s. These leaflets told civilians to evacuate and sometimes encouraged them to push their leaders to surrender. The "LeMay bombing leaflets," as they came to be known, were dropped on Hiroshima, Nagasaki and 33 other Japanese cities on August 1 and 3, 1945.

72   The Atomic Bomb in Images and Documents

USS *Indianapolis* (CA-35), off the Mare Island Navy Yard, California, July 10, 1945, after its final overhaul (Bureau of Ships Collection in the U.S. National Archives).

 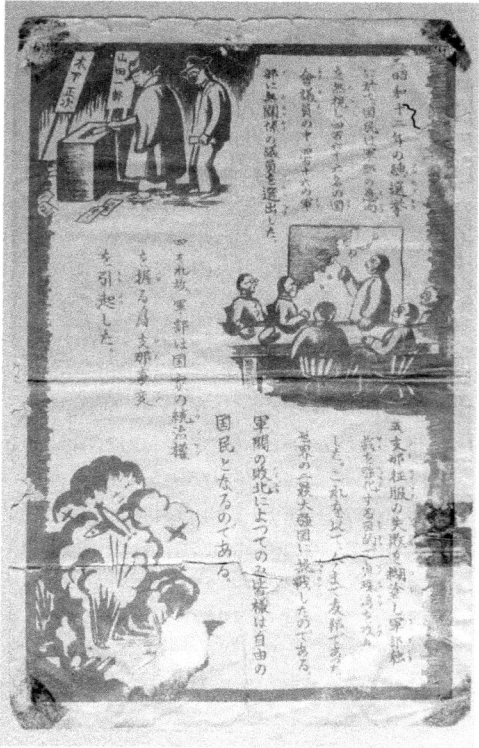

Leaflets dropped before atomic bombing of Hiroshima and Nagasaki (donated by Nakao Kelsou, Hiroshima Peace Memorial Museum).

Chapter 8. The Potsdam Declaration

Top and above: **Leaflets dropped before the atomic bombing of Hiroshima and Nagasaki (donated by Michinobu Okamura, Hiroshima Peace Memorial Museum).**

Top and above: **LeMay Leaflet. Counterclockwise, the text reads: Nagano, Takaoka, Kurume, Fukuyama, Toyama, Maizuru, Otsu, Nishinomiya, Maebashi, Koriyama, Hachioji, and Mito (donated by Kintsuchi Handa, Hiroshima Peace Memorial Museum).**

As documented in Richard Hubert's "The OWI Saipan Operation," the Japanese text on the reverse side of OWI #2106 carried the following warning:

> Read this carefully as it may save your life or the life of a relative or friend. In the next few days, some or all of the cities named on the reverse side will be destroyed by American bombs. These cities contain military installations and workshops or factories which

Chapter 8. The Potsdam Declaration

Top and above: **Leaflets dropped before atomic bombing of Hiroshima and Nagasaki (donated by Keiichi Muranaka, Hiroshima Peace Museum).**

produce military goods. We are determined to destroy all of the tools of the military clique which they are using to prolong this useless war. But, unfortunately, bombs have no eyes. So, in accordance with America's humanitarian policies, the American Air Force, which does not wish to injure innocent people, now gives you warning to evacuate the cities named and save your lives. America is not fighting the Japanese people but is fighting the military clique which has enslaved the Japanese people. The peace which America will bring will free the people from the oppression of the military clique and mean the emergence of a new and better Japan. You can restore peace by demanding new and good leaders who will end the war. We cannot promise that only these cities will be among those attacked but some or all of them will be, so heed this warning and evacuate these cities immediately.

OWI produced over 63 million leaflets that were dropped from B-29s as a warning to Japan.

Chapter 9

Planning the Atomic Bombing of Hiroshima

Starting in early August 1945, Colonel Tibbets sent three B-29s in formation, flying at 30,000 feet, over every primary and secondary target on the bombing list. This was done for four reasons. First, and most important, he wanted the Japanese to get used to seeing B-29s flying overhead without doing any bombing raids. Second, he wanted each crew to get experience in flying long distances and know exactly how to navigate to any target. Third, hundreds of photos would be taken at 30,000 feet to find suitable aiming points, and the pilots would become familiar with the landscape. Last, Tibbets wanted to know the exact weather conditions at any given target.

For each atomic mission, seven B-29s would take part. Each would leave Tinian at different intervals. Three would check on weather conditions at the primary and secondary targets. One would land in Iwo Jima for use in case the bomb-carrying plane ran into mechanical trouble. One plane would carry scientific instruments to measure the intensity of the blast, and another plane with photographic equipment would make a visual record of the explosion. The last plane would drop the bomb.

On August 2, 1945, Paul Tibbets and Tom Ferebee meet General LeMay at his headquarters in Guam. LeMay told Tibbets that Hiroshima was the primary target.

With historical maps and aerial photographs taken by the B-29s, Tibbets and Ferebee decided that the aiming point for the Hiroshima bombing should be a T-shaped bridge near the center of the city. The Japanese called it the Aioi Bridge.

On August 5, 1945, the weather conditions over Japan indicated clear weather, suitable for a visual bombing on August 6.

General LeMay issued Special Bomb Operational Order No. 13. The 509th Composite Group would carry out a Special Bomb Strike Mission against the primary target Hiroshima or against the secondary targets of Kokura or Nagasaki. The code name for the bombing was "Centerboard," and the mission was scheduled for August 6. Tibbets would have the code name "Dimples," and his plane, the *Enola Gay*, was "Dimples 82."

On the morning of Sunday, August 5, Tibbets asked a sign painter to place the lettering of *Enola Gay* on his plane. The painter followed Tibbets' instructions and painted the letters on the left side of the nose of the plane, just beneath the cockpit.

```
COMBAT CREWS OF THE 393RD BOMBARDMENT SQUADRON (VH)
OF THE 509TH COMPOSITE GROUP, BASED ON TINIAN, AND THE
B-29'S REGULARLY FLOWN (AND NAMED) BY THEM, JULY-AUGUST, 1945
```

Crew	Airplane Commander	B-29 Victor No.	Name	Serial Number
A-1	Ralph R. Taylor	83	Full House	44-27298
A-2	Edward M. Costello	95	Laggin' Dragon	44-86347
A-3	Ralph N. Devore	86	Next Objective	44-27299
A-4	Joseph E. Westover	73	Strange Cargo	44-27300
A-5	Thomas J. Classen (Gp Deputy CO) / Elbert B. Smith	94	(not named)	44-27346
B-6	John A. Wilson	71	Jabbit III	44-27303
B-7	James N. Price	84	Some Punkins	44-27296
B-8	Charles F. McKnight	72	Top Secret	44-27302
B-9	Paul W. Tibbets (Gp CO) / Robert A. Lewis	82	Enola Gay	44-27292
B-10	George W. Marquardt	88	Up an' Atom	44-27304
C-11	Claude R. Eatherly	85	Straight Flush	44-27301
C-12	Herman S. Zahn	90	Big Stink	44-27354
C-13	Frederick C. Bock	77	Bockscar	44-27297
C-14	Norman W. Ray	91	Necessary Evil	44-27291
C-15	Charles W. Sweeney (Squadron CO) / Charles D. Albury	89	Great Artiste	44-27353

```
NOTES: The table is based on documents available to me plus inputs
from several 393rd members (F. Bock, April 1990).

    Unless circumstances dictated otherwise, a crew flew missions
in its "own" airplane. Exceptions on August 6 and 9 are as
follows.
    V-91 was the camera plane on the Hiroshima mission, flown by
B-10.
    V-77 was the bomb plane on the Nagasaki mission, flown by C-15.
    V-89 was the instrument plane on the Nagasaki mission, flown
by C-13.
    V-90 was the camera plane on the Nagasaki mission, flown by
C-14 with James I. Hopkins (Gp Op. Officer) as AC in place of Ray.
```

Combat crews of the 393rd Bombardment Squadron (author's collection).

In a letter to this author dated May 16, 1978, Tibbets explained why pilots give names to their aircraft and, in particular, why he chose the name *Enola Gay*:

> During the war hundreds of aircraft carried names of individuals (mostly wives and sweethearts), cities or names intended to set that airplane or crew off from the rest. Custom was not involved; it was a matter of inclination on the part of the aircraft commander and/or crew. I was aware that my aircraft would "live in history." Consequently, I wanted its name to be unoffensive and one unlikely of duplication. Since my mother was the only member of my family who did not try to persuade me to "give up" flying, and since her name was most unlikely of duplication—need I say more?

In the afternoon, Tibbets had the tail insignias of the seven planes that were to participate in the Hiroshima mission repainted. The encircled arrow insignia of the 509th Composite Group was replaced by an encircled "R," which was used by the other B-29s of the 20th Air Force at North Field on Tinian.

"Little Boy" was loaded onto a trailer and slowly moved into a loading pit. The *Enola Gay* (No. 82) was towed to the pit. "Little Boy" was then secured into

EXECUTIVE JET AVIATION, INC.
P.O. BOX 19707 · COLUMBUS, OHIO 43219 · 614-237-0363 · TELEX 24-5495

PAUL W. TIBBETS
PRESIDENT

May 16, 1978

Mr. Samuel S. Kloda
5817 Eldridge Avenue
Cote St. Luc., Quebec
Canada H4W 2E3

Dear Mr. Kloda:

A quick reply to your questions of May 11th.

1) During the war hundreds of aircraft carried names of individuals (mostly wives and sweethearts), cities or names intended to set that airplane or crew off from the rest. Custom was not involved; it was a matter of inclination on the part of the airplane commander and/or crew. I was aware that my aircraft would "live in history", consequently, I wanted its name to be unoffensive and one unlikely of duplication. Since my mother was the only member of my family who did not try to persuade me to "give up" flying, and since her name was most unlikely of duplication--need I say more?

2) Every crew in the theater posed for pictures immediately after their arrival from the U.S. The purpose was for IDENTIFICATION. Contrary to what the pilots you have spoken to have had to say, it is the first I have heard about such a thing being "unlucky".

3) I can not confirm or deny as to anything being written on the bomb. Nothing was painted in a contrasting color. I have been told, and accepted statements that names, comments or slogans were written in pencil or pen by the loading crew.

I have just completed "my story" which is being published by Stein & Day and is to be released this August.

Sincerely,

Paul W. Tibbets

PWT:v

General Paul Tibbets answers questions posed by Samuel S. Kloda, dealing with names of airplanes (author's collection).

the bomb bay, and all the pull-out wires were attached to their hooks in the plane.

Captain William Parsons and Lieutenant Morris Jeppson climbed into the bomb bay and spent around two hours practicing the propellant loading operation. The bomb bay doors were closed, and guards posted until takeoff time.

Seven B-29s of the 393rd Bombardment Squadron were assigned to participate in the bombing mission.

On the morning of August 5, 1945, Paul Tibbets asked the painting department to paint the name *Enola Gay* on his plane. In the afternoon, he also asked that all seven planes making the trip to Hiroshima be painted with the letter "R" replacing the original arrow. Autographed by Morris Jeppson, weapons test officer (U.S. Air Force; author's collection).

Rare color photograph of *Enola Gay*, just before the repainting of the arrow and circle on the tail and replacing it with a big black "R," also inside a circle. Autographed by George Caron (photograph by George Caron; author's collection).

Chapter 9. Planning the Atomic Bombing of Hiroshima 81

Enola Gay, parked before its painting job. Autographed by Tom Ferebee, Paul Tibbets, and Theodore Van Kirk (U.S. Air Force; author's collection).

Bockscar before painting. Autographed by Fred Bock and Fred Olivi (U.S. Air Force; author's collection).

Bockscar parked before painting. Autographed by Charles Albury, co-pilot (U.S. Air Force; author's collection).

Three B-29s would take off one hour before the strike plane and report on the weather conditions at each of the three targets. *Straight Flush*, piloted by Major Claude Eatherly, was to fly to Hiroshima, while *Jabbitt III*, piloted by Major John Wilson, flew to Kokura, and *Full House*, piloted by Major Ralph Taylor, went to Nagasaki. *Top Secret*, flown by Charles McKnight, would land on Iwo Jima and stand by in case the *Enola Gay* ran into problems.

Later that morning, the *Enola Gay*, piloted by Colonel Paul Tibbets, would fly to Iwo Jima. Major Charles Sweeney, flying the *Great Artiste*, and Captain George Marquardt, flying *No. 91*, would follow Tibbets to Iwo Jima and then to the selected target.

At the chosen target, Sweeney would drop blast gauges and canisters, suspended from parachutes, to measure the blast output of the bomb. The statistical information would then be relayed by radio telemetry to recorders in the plane. *No. 91*, equipped with various cameras, would record the photographic evidence of the historic mission.

Takeoff time was set for 2:45 a.m. the following day.

Opposite, bottom: **Captain William S. Parsons was an American naval officer who worked as an ordnance expert on the Manhattan Project during World War II. He was the weaponeer on the *Enola Gay*, the aircraft that dropped the "Little Boy" atomic bomb on Hiroshima, Japan, August 6, 1945 (Air Force Historical Research Agency).**

Chapter 9. Planning the Atomic Bombing of Hiroshima

With its new paint job completed, the *Enola Gay* was backed over the pit to up-load the "Little Boy" atomic bomb (Air Force Historical Research Agency).

Flight crew of *Enola Gay*. *Left to right kneeling:* Technical Sergeant George R. Caron, Sergeant Joe S. Stiborik, Staff Sergeant Wyatt E. Duzenbury, Private First Class Richard H. Nelson, Sergeant Robert H. Shumard. *Left to right standing:* Major Thomas W. Ferebee, group bombardier; Captain Theodore J. Van Kirk, navigator; Colonel Paul W. Tibbets, 509th Composite Group commanding officer and pilot; Captain Robert A. Lewis, airplane commander. This entire crew made the historical flight over Hiroshima (missing is Lt. Jacob Beser). Autographed by George Caron (U.S. Air Force; author's collection).

Just after evening mess, there was a pre-flight briefing of all seven crews that would be taking part in this historical event. The briefing in the assembly hut involved instructions regarding the route that would be followed, the radio frequencies to be used, and current weather conditions.

The crews were also given locations of rescue ships and submarines that would be in the flight path of the B-29s, in case of a forced ditch or enemy attack. Each person wore underneath his coveralls a survival vest containing a drinking-water kit, first-aid package, fishhooks, and emergency food rations. All crew members were given handguns and were wearing them on the flight.

The U.S. Navy issued an order that none of its ships were to be within a 50-mile radius of the prime targets.

The briefing closed with a prayer, made by Chaplain William B. Downey, for a safe return of the crews and a quick end to the war.

The only important change was on the first leg of the flight to Iwo Jima. Instead of an altitude of 6,000 feet, Tibbets would fly at 5,000 feet. This approach

Chapter 9. Planning the Atomic Bombing of Hiroshima

Chaplain William Downey (right) with Samuel S. Kloda. Downey was the chaplain to the 509th Composite Group (author's collection).

would allow Captain Parsons enough time to arm the bomb in the unpressurized bomb bay.

At 11:00 p.m., in the combat crew lounge, there was a final briefing for the crews that would accompany the *Enola Gay* to Iwo Jima and then to the strike target. The briefing was conducted by Colonel Tibbets, Captain Parsons, and Professor Ramsey. Tibbets went over all the details of the flight, including the codes to be used, radio frequencies, rendezvous points, and altitudes.

Personnel from Project Alberta were assigned to fly on the two observation aircrafts. Harold M. Agnew, Luis W. Alvarez, and Lawrence H. Johnston would be on the *Great Artiste*, while Bernard Waldman was assigned to *No. 91*. Waldman had conducted blast measurements for the Trinity nuclear test and had served on Tinian. He would be using a fast-acting camera on board *No. 91*.

The three members of Project Alberta on board the *Great Artiste* would immediately drop the "Bangometer" canisters to measure the force of the blast. At Los Alamos, Alvarez had worked on the design of explosive lenses and the development of exploding-bridgewire detonators. Agnew had worked on the Cockcroft-Walton generator. Alvarez and Johnston measured the blast effect of the "Little Boy" bomb.

Tibbets' final statement was taken down by William L. Laurence, a *New York*

WILLIAM B. DOWNEY
833 LAKESHORE DRIVE
MOUNT DORA, FLORIDA 32757

27 November

Samuel S. Kloda, B.Sc., M.Sc.
5817 Eldridge Ave.,
Cote St. Luc, Quebec
Canada, H4W 2E3

Dear Samuel

Frequently when I go somewhere to a meeting, a convention, reunion, whatever, I will meet someone who stands out, whom I will remember with appreciation & with good memories — and from Wendover 1990, you are the guy. It was a pleasure to meet with you, come to know you.

Relative to the 509th Composite Group, don't become too enchanted with it. They are an accident of history. Too many who were in the war in some notable don't yet find the meaning of their life in their military juxtaposition. Find rather one's meaning in being a child of God as indeed your namesake found his meaning in being the mouth of God in his time.

Being a part of the 509th was an accidental honor for me, but I am far more proud of having built a large parish of people who gather yet around the Word of God, or starting a program of the care of the aged in the Milwaukee area, including in which program is an

Above and opposite: Letter from Chaplain William Downey to Samuel S. Kloda (author's collection).

Times science writer who was given permission by the U.S. government to write the official history of the atomic bomb development. (Later, he would win the Pulitzer Prize for his vivid description of the atomic bombings.) This is what was recorded by Laurence on his notepad:

institution which now in this latter time cares for the needs of some six hundred plus persons. The investment of life in response to some need of man is significant. And who can tell what shall remain and what shall pass away? Remember your namesake, Samuel, who in a real sense made Israel possible as he anointed David and out of this came great and mighty things.

This is the curse of a preacher: he starts to preach when no one asked him to. Enough.

Samuel, it was a pleasure to meet you. I hope all goes well with you. Now as legions gather in the lands through which the Jews once moved, let us pray that peace will prevail, evil be overthrown. Shalom aleichem

Sue Donner

P.S. Thank you for the pictures you sent to me. I hope I have returned to you all you wanted returned.

Go with God!

> Tonight is the night we have all been waiting for. Our long months of training are to be put to the test. We will soon know if we have been successful or failed. Upon our efforts tonight it is possible that history will be made. We are going on a mission to drop a bomb different from any you have ever seen or heard about. This bomb contains a destructive force equivalent to twenty thousand tons of TNT.

Professor Ramsey and Captain Parsons confirmed the tremendous power of the bomb. Parsons is quoted as saying, "The bomb you are going to drop is something new in the history of warfare. It will be the most destructive weapon ever devised. We think it will wipe out almost everything within a three-mile area, maybe slightly more, maybe somewhat less."

Tinian, Mariana Islands. Members of the 509th Composite Group attend a briefing conducted by Captain Joseph Buscher and Lieutenant Charles Levy in preparation for the first atomic bombing on Hiroshima, Japan, August 6, 1945. Captain William "Deak" Parsons points to map photographs. Autographed by Jacob Beser (U.S. Air Force; author's collection).

Finally, Tibbets reminded everyone that immediately after release of the bomb, the *Enola Gay* would make a sharp 155-degree turn to the right, while Chuck Sweeney's *Great Artiste* would make a 155-degree turn to the left. After making the turn, Sweeney would immediately release the instrument canisters by parachute to record the strength of the shock waves and the amount of radioactivity in the air. George Marquardt, flying *No. 91*, would lag far enough behind to avoid the shock waves and would record photographic evidence of the bombing.

Each member of these three crews was given special goggles with an adjustable polaroid lens. These were to be worn as soon as the bomb was released, so as to protect their eyes from the bomb flash, which was estimated to be ten times brighter than the sun.

On exiting the crew lounge, Beser was handed a small piece of rice paper by scientist Ed Doll. On it were the four radio frequencies that the bomb's radar (proximity fuzz) would use to measure the distance from the ground as it descended. The rice paper was to be swallowed if Beser was in danger of being captured.

After the final briefing, the three crews went for a special breakfast at 12:30 a.m. As they were entering the mess hall, the crews of the three weather

Chapter 9. Planning the Atomic Bombing of Hiroshima

Tinian, Mariana Islands. Members of the 509th Composite Group make plans for the first atomic bomb mission against Hiroshima on August 6, 1945. Autographed by Paul Tibbets (U.S. Air Force; author's collection).

planes—*Straight Flush*, *Jabbitt III*, and *Full House*—were leaving. Their flight time was 1:37 a.m. Each would fly to its designated target city and confirm by radio whether the weather conditions were suitable for a visual bomb drop.

As Tibbets and the other crew members were leaving the mess hall, Tibbets was handed a small pillbox containing 12 cyanide capsules. Don Young, the flight surgeon, had discussed this subject before with Tibbets. There was documented proof of atrocities inflicted by the Japanese against prisoners of war. In the unlikely event that the *Enola Gay* was forced to land or the crew had to bail out after dropping the bomb, each crew member would be given a pill. It was up to each member to decide on his own whether to take the pill.

Tibbets walked around the *Enola Gay*, checking for any oil leaks and making sure the tires were fully inflated.

Before the flight, there were official Army photographers who took group photos of the entire crew and then each man separately in front of the plane. Photographers wanted photos of Tibbets waving from the open cockpit, showing the name *Enola Gay*.

This is the list of the 12 crew members of the *Enola Gay* and their functions:

Colonel Paul Tibbets: pilot and aircraft commander
Captain Robert A. Lewis: co-pilot and aircraft commander
Major Thomas Ferebee: bombardier
Captain Theodore "Dutch" Van Kirk: navigator
Staff Sergeant Wyatt E. Duzenbury: flight engineer
First Lieutenant Jacob Beser: radio countermeasures (only person to fly on the strike plane on both atomic bombing missions)

Chapter 9. Planning the Atomic Bombing of Hiroshima

George Marquardt (right) with Samuel S. Kloda. Marquardt flew the observer plane on the Hiroshima flight (author's collection).

Opposite: Los Alamos scientists holding one of the instruments used to measure the effects of the first atomic bomb drop on Japan. *Top row, left to right:* Harold M. Agnew, Luis Alvarez. *Bottom, left to right:* Lawrence Johnston, Bernard Waldman.

Harold M. Agnew was an American physicist. He was a scientific observer on the Hiroshima bombing mission and, later, the third director of the Los Alamos National Laboratory.

Luis Alvarez was an American experimental physicist and inventor. Alvarez worked at the University of Chicago on nuclear reactors for Enrico Fermi before coming to Los Alamos to work for Robert Oppenheimer on the Manhattan Project. Alvarez worked on the design of explosive lenses and the development of exploding-bridgewire detonators. As a member of Project Alberta, he observed the Trinity nuclear test from a B-29 Superfortress and later the bombing of Hiroshima from the *Great Artiste*. He was awarded the Nobel Prize in Physics in 1968 for development of the hydrogen bubble chamber, which enabled the discovery of resonance states in particle physics.

Lawrence Johnston was the only man to witness all three atomic explosions in 1945: the Trinity nuclear test and the atomic bombing of Hiroshima and Nagasaki. At the MIT Radiation Laboratory, he invented ground-controlled approach radar. In 1944, he went to the Manhattan Project's Los Alamos Laboratory, where he invented the exploding-bridgewire detonator.

Bernard Waldman was an American physicist who flew on the Hiroshima atomic bombing mission as a cameraman during World War II. At Los Alamos, he headed a group that conducted blast measurements for the Trinity nuclear test, and he later served on Tinian with Project Alberta. Autographed by Harold Agnew (Los Alamos Photographic Laboratory; author's collection).

James Van Pelt (left) with Samuel S. Kloda. Van Pelt was the navigator on the *Bockscar*. Photograph taken in front of the 509th Composite Group Memorial in Wendover, Utah (author's collection).

Theodore Van Kirk (left) with Samuel S. Kloda. Van Kirk was the navigator on the *Enola Gay* (author's collection).

Chapter 9. Planning the Atomic Bombing of Hiroshima

Colonel Paul W. Tibbets waves from his cockpit before takeoff on August 6, 1945. Autographed by Paul Tibbets, Theodore Van Kirk, Tom Ferebee, and Jacob Beser (U.S. Air Force; author's collection).

    Technical Sergeant George "Bob" R. Caron: tail gunner
    Sergeant Joseph S. Stiborik: radio operator
    Sergeant Robert H. Shumard: assistant flight engineer
    Private First Class Richard H. Nelson: VHF radio operator
    Captain William "Deak" Parsons: weaponeer and mission commander
    Second Lieutenant Morris R. Jeppson: assistant weaponeer

Chapter 10

The Atomic Bombing of Hiroshima

"Little Boy" had 64 kilograms of uranium (U-235). It was a gun-type fission weapon. Fission would be accomplished by shooting a hollow cylinder of enriched uranium (the "bullet") onto a solid cylinder of the same material (the "target") by means of a charge of nitrocellulose propellant powder.

A proximity fuse is a fuse that detonates an explosive device automatically when the distance to the target becomes smaller than a predetermined value. The device uses a small, short-range Doppler radar.

The proximity fuse was one of the most important technological innovations of World War II. Its secret technology was guarded to the same extent as the atomic bomb. It is a miniature radar set. Its built-in transmitter would send out radio waves that would hit the ground, bounce back, and be received by the bomb's radar antennae. At the predetermined altitude, the radar was set to close a switch, which would start the bomb mechanism.

"Little Boy" had four tail fans to keep it in a vertical position once it was released from the bomb bay. This was necessary because then the proximity fuse would know exactly when to trigger detonation.

In *The Tibbets Story*, Paul Tibbets writes, "A number of our people stopped by to scrawl messages on the side of the bomb. One, addressed to Emperor Hirohito, was signed 'From the Boys of the Indianapolis,' implying revenge for the loss of those aboard the cruiser that had been sunk after delivering the material for this bomb."

On August 5, 1945, four B-29s had crashed and burned on the runways at Tinian. This was especially dangerous because a crash of a B-29 that was carrying a "live" bomb could set off a fire and detonate the bomb, which would wipe out most of Tinian.

Captain William Parsons, the weaponeer and mission commander, decided to load the four cordite powder bags into the gun breech to arm the bomb in flight. Captain Parsons, as mission commander, was the personal representative of General Groves.

On August 6, 1945, at 2:45 a.m., Tibbets started the four 2,200-horsepower Wright Cyclone engines of the *Enola Gay*, and then he accelerated down the 8,500-foot runway. The plane was heavily loaded with 7,000 gallons of fuel and a bomb weighing 9,000 pounds. Tibbets held the nose wheel down until the

The "Little Boy" uranium bomb detonated at 1,890 feet over Hiroshima, Japan. The bomb was 28 inches in diameter and 120 inches long. It weighed about 9,700 pounds and had a yield equivalent to approximately 15 kilotons of TNT. Signed by Technical Sergeant George Robert "Bob" Caron (tail gunner), Joseph A. Badali (bomb assembler), Richard Nelson (radio operator) (Los Alamos National Laboratory; author's collection).

"Little Boy" replica. Autographed by Paul Tibbets (U.S. Air Force; author's collection).

Colonel Paul Tibbets starts his historic flight down the runway on Tinian. Autographed by Thomas Ferebee (U.S. Air Force; author's collection).

last possible moment, using nearly the whole runway, taking off at 155 miles per hour.

The *Great Artiste* (flown by Charles Sweeney) and *No. 91* (flown by George Marquardt) subsequently left Tinian at two-minute intervals to accompany the *Enola Gay*. The three planes followed a north-northwest course to Iwo Jima.

Arming the bomb started eight minutes after takeoff, with *Enola Gay* cruising at an altitude of 4,700 feet. Parsons, assisted by Morris Jeppson, began the difficult task of inserting gunpowder and the conventional explosive charge into the core of "Little Boy." Parsons disconnected the primer wires, removed the breech plug, inserted the powder bags, replaced the breech plug, and reconnected the wires. The weapon was now ready, except for a safety precaution. Jeppson would later return to the bomb bay to exchange the three safety plugs between the electrical connectors of the internal battery and the firing mechanism from green to red. Then the bomb would be fully armed.

After three hours of flying time, the *Enola Gay* climbed to 9,300 feet. At 5:55 a.m., the *Enola Gay* arrived at Iwo Jima. While circling over Iwo Jima, the *Great Artiste* and *No. 91* took positions behind each wing of *Enola Gay* for their trip to Japan. Charles McKnight, flying *Top Secret*, landed his plane at Iwo Jima.

At 6:00 a.m., Jeppson returned to the bomb bay and unscrewed the three green plugs, replacing them with red ones. Parsons reported to Tibbets that the bomb was ready.

Chapter 10. The Atomic Bombing of Hiroshima

Preflight photo. *Left to right, standing:* Lieutenant Colonel John Porter, ground maintenance officer on B-29 *Enola Gay*; Captain Theodore J. Van Kirk, navigator; Major Thomas W. Ferebee, bombardier; Colonel Paul W. Tibbets, 509th Composite Group commander and pilot; Captain Robert A. Lewis, co-pilot; and Lieutenant Jacob Beser, radar countermeasures officer. *Front row, left to right:* Sergeant Joseph S. Stiborik, radar operator; Technical Sergeant George R. Caron, tail gunner; Private First Class Richard H. Nelson, radio operator; Sergeant Robert H. Shumard, assistant engineer; and Staff Sergeant Wyatt E. Duzenbury, flight engineer. Missing from picture: Navy Captain William S. Parsons and Lieutenant Morris Jeppson. Autographed by Bob Lewis (U.S. Air Force; author's collection).

Tibbets switched on the intercom and boldly told the crew, "We are carrying the world's first atomic bomb."

At 6:07 a.m., all three planes headed for Japan. From Iwo Jima, the aircraft passed over the western part of Shikoku, west of Kure, and then straight toward Hiroshima.

At 7:30 a.m., Parsons made necessary adjustments on the console that controlled the bomb's circuitry. He informed Tibbets that "Little Boy" was now "alive."

At 7:41 a.m., Tibbets started his climb to an altitude of 30,700 feet.

Tibbets had orders that he should have a visual bombing run, so that photographic evidence could be gathered and distributed to the world. Otherwise, no bomb should be dropped.

At 8:30 a.m., the *Enola Gay* climbed to 32,700 feet. It received a coded message from Major Claude Eatherly's *Straight Flush*, flying over Hiroshima: the cloud cover was less than three-tenths at all altitudes, and it was clear to bomb target.

Tibbets sent a message to Sweeney and Marquardt that Hiroshima was the primary target.

In 1945, Hiroshima was a city with approximately 300,000 civilians, 43,000 soldiers, and 45,000 Korean forced laborers. It was a waterfront city with a large port. It was also one of Japan's largest military depots, with 25,000 soldiers of the Second Army.

The aiming point, or the initial point (as known to the bombardier), was 15½ miles east of the center of the city. As the *Enola Gay* approached the aiming point, Tibbets reminded the crew to put on their special goggles when they received a message from Tom Ferebee that the bomb had been dropped. The same message was also sent in code, by radio, to Sweeney's *Great Artiste* and Marquardt's *No. 91*. (Later the *No. 91* crew would rename their plane *Necessary Evil*.)

There was no opposition from any Japanese aircraft, nor was there any anti-aircraft fire.

As the *Enola Gay* approached Hiroshima, there was a wind velocity of eight

Pre-strike aerial view of Hiroshima showing the high density of the built-up area. Fire lanes cleared by the Japanese are plainly visible (U.S. Air Force; author's collection).

Total area devastation by the atomic bomb strike on Hiroshima. Autographed by Jacob Beser (U.S. Air Force; author's collection).

knots. To compensate, the ground speed was reduced to 285 knots (330 mph), and the drift to the right required an 8-degree correction.

Ferebee, who was an expert bombardier, adjusted the Norden bombsight according to his calculations, knowing what the crosswind would do to the falling bomb.

Thomas Ferebee (right) with Samuel S. Kloda. Ferebee was the bombardier on the *Enola Gay* (author's collection).

At 10 miles out, Ferebee found his aiming point, the T-shaped Aioi Bridge, which was easily visible. Then Tibbets put the plane in autopilot; now Ferebee had control of the aircraft. The sighting and release mechanism were synchronized, so that the bomb drop would take place at a predetermined point in the bomb run.

In an interview with the Pulitzer Prize winner Louis "Studs" Terkel (*The Guardian*, August 6, 2002), Tibbets described the last minute before "Little Boy" was released:

> The airplane has a bomb sight connected to the autopilot and the bombardier puts figures in there for where he wants to be when he drops the weapon, and that's transmitted to the airplane. We always took into account what would happen if we had a failure and the bomb bay doors didn't open: we had a manual release put in each airplane so it was right down by

Chapter 10. The Atomic Bombing of Hiroshima

the bombardier and he could pull on that. And the guys in the airplanes that followed us to drop the instruments needed to know when it was going to go. We were told not to use the radio, but, hell, I had to. I told them I would say, "One minute out," "Thirty seconds out," "Twenty seconds" and "Ten" and then I'd count, "Nine, eight, seven, six, five, four seconds," which would give them a time to drop their cargo. They knew what was going on because they knew where we were. And that's exactly the way it worked, it was absolutely perfect.

At 9:14:17 a.m., just 60 seconds before the automatic bomb release, on the earphones of all crew members of the three planes, there was a high-pitched radio tone indicating that the bomb would drop in exactly one minute. It was a signal for them to put on their goggles and for Tibbets and Sweeney to get ready to make their 155-degree turns.

At 9:15:17 a.m., the doors of the bomb bay opened and "Little Boy" started his descent. Soon the four stabilizing fins would force the nose in a downward position. The proximity fuse would then be activated.

At exactly the same moment, the *Enola Gay* jolted upward. The plane was now 9,000 pounds lighter.

Now, Tibbets made his often-practiced escape maneuver, a 155-degree diving turn to the right, away from the aiming point. Sweeney made the same sharp 155-degree turn to the left.

The four flaps of "Little Boy" opened, and the now vertical bomb accelerated until it reached 1,890 feet above the ground.

In a letter to General Leslie Groves, Dr. Oppenheimer indicated that the optimum height for detonation was at 1,910 feet. Also, detonating an atomic bomb in the air, at such an altitude, causes the blast to have a larger geographical impact. The city of Hiroshima was completely flat. The atomic bomb blast would expand unhindered in all directions at unbelievable speed and force.

"Little Boy" exploded 900 feet from its aiming point.

In a millisecond, the chain reaction started an expanding fireball. The fireball travelled at 100 times the speed of sound.

George Caron, in the tail of the *Enola Gay*, was the only crew member to see the fireball. At the moment of nuclear fission, the temperature was calculated to be 100 million degrees Fahrenheit.

Caron took several photographs using a handheld K-20 aerial camera. On his return, these photos would be on the front pages of every newspaper in the world.

As envisioned, the first shock wave hit the *Enola Gay* in 60 seconds. Tibbets estimated that the initial shock wave struck the plane about nine miles from the point of explosion. Two more shock waves hit the *Enola Gay* at about 15 miles from the explosion, though with much less impact. Tibbets would later describe the first violent shock: "I found the effect to be much like that produced by an anti-aircraft shell exploding near a plane."

Tibbets announced over the intercom, "Fellows, you have just dropped the first Atomic bomb in history."

Tibbets circled to view the rising mushroom-shaped cloud that seemed to be coming toward the plane. The cloud now had climbed to 70,000 feet, sucking up everything on the ground.

George Caron (right) with Samuel S. Kloda. Caron was the tail gunner on the *Enola Gay* (author's collection).

First photograph of Hiroshima taken by George Caron from the turret of *Enola Gay* (Library of Congress).

Chapter 10. The Atomic Bombing of Hiroshima

At the time this photograph was taken, smoke billowed 20,000 feet above Hiroshima while smoke from the burst had spread over 10,000 feet from the target at the base of the rising column. As tail gunner, seated in the rear of the B-29, George R. Caron was the first to see the atomic bomb mushroom cloud. Caron had been given a camera and took a roll of photographs as the mushroom cloud ascended. His photographs of the explosion were printed on millions of leaflets that were dropped over Japan the next day. Autographed by George Caron (U.S. Air Force; author's collection).

Clouds and smoke obscured Hiroshima from reconnaissance planes seeking to photograph atomic bomb damage after the target was hit by the Army Air Force. This picture was taken from 25,000 feet in elevation by the photographic plane. Autographed by Bob Lewis, co-pilot (U.S. Air Force; author's collection).

Tibbets later would describe a devastating scene:

> The giant purple mushroom, which the tail-gunner had described, had already risen to a height of 45,000 feet, three miles above our own altitude, and was still boiling upward like something terribly alive. It was a frightening sight, and even though we were several miles away, it gave the appearance of something that was about to engulf us.

Chapter 10. The Atomic Bombing of Hiroshima

> August 6, 1945 at 2:27 A.M. We started engines. I think everyone will be relieved when we have left our bomb with the Japs and get halfway home, or better still, all the way home. 7:30 A.M.....We are loaded. The bomb is now alive and it's a funny feeling knowing it's right in back of you ... There'll be a short intermission while we bomb our target ... If I live a hundred years I'll never quite get those few minutes out of my mind ... I5 seconds after the flash, there were two very distinct slaps ... and there in front of our eyes was without a doubt the greatest explosion man has ever witnessed.
>
> *Capt Bob Lewis*

Bob Lewis kept a log during his flight to Hiroshima. He printed the major part and signed it for me (author's collection).

Even more fearsome was the sight on the ground below. At the base of the cloud, fires were springing up everywhere amid a turbulent mass of smoke that had the appearance of bubbling hot tar. If Dante had been with us in the plane, he would have been terrified! The city we had seen so clearly in the sunlight a few minutes before was now an ugly smudge. It had completely disappeared under this awful blanket of smoke and fire.

A feeling of shock and horror swept over all of us.

Tail gunner Caron was the only crew member to actually witness the explosion. As he would later recall in Andrew Rotter's book, *Hiroshima: The World's Bomb*:

> A column of smoke is raising fast. It has a fiery red core. A bubbling mass, purple, grey in color with a red core. It's like a mass of bubbling molasses. The mushroom is spreading out. It's maybe a mile or two wide and half a mile high. It's growing up and up and up. It's nearly level with us and climbing. It's very black, but there is a purplish tint to the cloud. The base of the mushroom looks like a heavy undercast that is shot through with flames.

Michael Bess, in *Choices Under Fire: Moral Dimensions of World War II*, quotes a college history professor:

> I climbed Hikiyama Hill and looked down. I saw that Hiroshima had disappeared.... I was shocked by the sight.... What I felt then and still feel now I just can't explain with words. Of course I saw many dreadful scenes after that—but that experience, looking down and

Mushroom cloud over Hiroshima (U.S. Army photo, from the Hiroshima Peace Memorial Museum).

> finding nothing left of Hiroshima—was so shocking that I simply can't express what I felt.... Hiroshima didn't exist—that was mainly what I saw—Hiroshima just didn't exist.

Medical doctor Michihiko Hachiya was an eyewitness to the Hiroshima bombing. He wrote in his book *Hiroshima Diary*:

> Nothing remained except a few buildings of reinforced concrete.... For acres and acres the city was like a desert except for scattered piles of brick and roof tile. I had to revise my meaning of the word destruction or choose some other word to describe what I saw. Devastation may be a better word, but really, I know of no word or words to describe the view.

It was estimated that the bomb blast had a yield equivalent to 15,000 tons (15 kilotons) of TNT.

"Little Boy" exploded a few hundred feet from the Aioi Bridge.

In one millionth of a second, at least 40,000 people were incinerated. Their bodies became small lumps of charcoal. It was estimated that at least 80,000 civilians and soldiers died immediately, mostly vaporized. Later another 70,000 would die of injuries and radiation poisoning. Five years later, the death toll would reach 200,000.

Within a one-mile area of the bomb blast, every person and every building vanished. Outside this radius, tens of thousands experienced flash burns and severe cuts caused by flying shards and exploding glass thundering through the thick black smoke.

For most of the survivors, purple spots appeared on their bodies, their hair

Chapter 10. The Atomic Bombing of Hiroshima

Photographed from *No. 91* airplane minutes after the detonation. The pilot was Captain George Marquardt. Autographed by Paul Tibbets (U.S. Air Force; author's collection).

fell out, and they developed infections, torn skin, and swollen and bleeding gums. Later, cancer would set in. The survivors always lived in constant fear of further illness and death.

The blast traveled at two miles per second, with raging heat and particles of radiation.

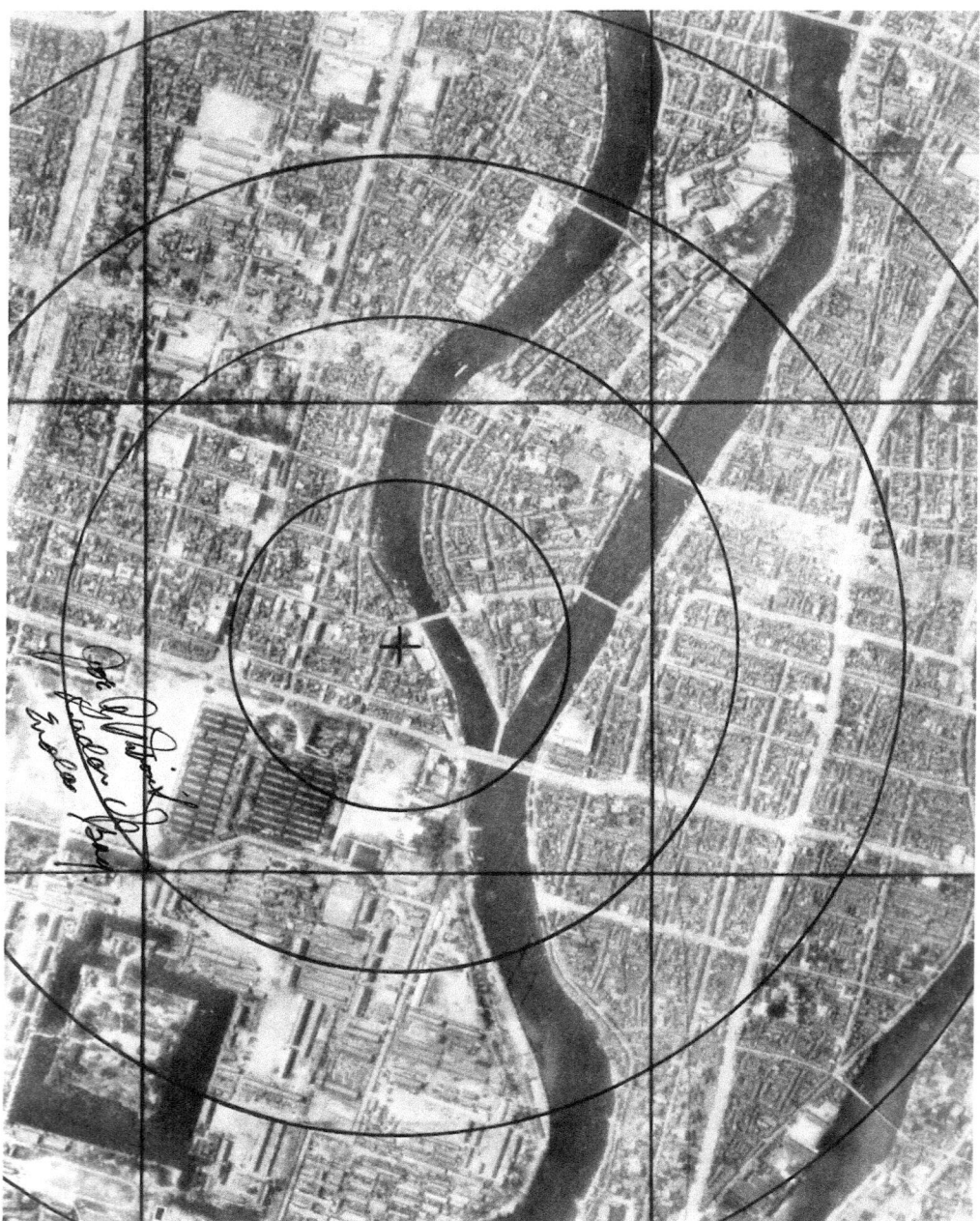

"Ground Zero" is the spot directly below the explosion of the bomb. In this photograph, there are 1,000-feet concentric circles around the aiming point. Autographed by Joseph Stiborik, radio operator (U.S. Air Force; author's collection).

Opposite, bottom: General panoramic view of Hiroshima after the atomic bombing on August 6, 1945. This picture shows the devastation 0.4 miles from the aiming point. Autographed by Charles Sweeney (U.S. Army; author's collection).

View of a Hiroshima bridge, 4,400 feet east of bomb explosion. Autographed by Theodore Van Kirk (U.S. Army; author's collection).

Hiroshima Prefectural Industrial Promotion Hall (now Atomic Bomb Dome) reduced to rubble. Location: Peace Memorial Park. Vicinity distance from hypocenter: 260 meters (U.S. Army photo, from Hiroshima Peace Memorial Museum).

Thousands who fled the extreme heat and flames ran to the port, believing that this would save them. Unfortunately, it did not. The heat from the explosion had superheated the water. Those who sought water as a refuge were immediately boiled to death.

Although the Hiroshima bomb proved to have unbelievable destructive power, scientists calculated its overall efficiency was only 1 percent.

As Tibbets and the crews headed home, they hoped that this mission would end the war. From over 250 miles away, at an altitude of 26,000 feet, the mushroom cloud could still be seen by Caron.

Parsons sent a coded message to General Farrell on Tinian: "82 V 670, Able, Line 1, Line 2, Line 6, Line 9." Deciphered, it read, "Clear cut, successful in all respects, Visual effects greater than Trinity, Hiroshima, Conditions normal in airplane following delivery, proceeding to regular base."

The *Enola Gay* returned to North Field, Tinian, at 2:58 p.m., 12 hours after its initial departure. The crew members were met with a hero's welcome. Photographers and movie cameras were everywhere.

The welcoming group consisted of more than 200 officers and enlisted men, including General Carl Spaatz, commander of the Strategic Air Force; General

Chapter 10. The Atomic Bombing of Hiroshima

This photograph was taken from the roof of the Chugoku Shimbun building. Location: Kami-Nagarekawa-cho (now Ebisu-cho). Distance from hypocenter: approximately 870 meters. One could see the entire city, the Chugoku mountain range and surrounding islands. That entire landscape was transformed on August 6, 1945 (photograph by Shigeo Hayashi, from Hiroshima Peace Memorial Museum).

Aerial view of Hiroshima, showing atomic damage. From an album by the Allied Reparations Committee, headed by Ambassador Edwin W. Pauley, 1945–1947 (Harry S. Truman Library & Museum).

This ferro-concrete chimney was part of a soy sauce brewery; somehow it managed to withstand the blast. The chimney is standing at the lower right of the photograph. Location: 900 meters from the hypocenter, Chigiya-cho (now Horikawa-cho) (photograph by Takashi Saeki, from Hiroshima Peace Memorial Museum).

Chapter 10. The Atomic Bombing of Hiroshima

This photograph was taken from the roof of the Chugoku Shimbun building. Location: Kami-Nagarekawa-cho (now Ebisu-cho). Distance from hypocenter: approximately 870 meters. One could see the entire city, the Chugoku mountain range and surrounding islands. That entire landscape was transformed on August 6, 1945 (photograph by Shigeo Hayashi, from Hiroshima Peace Memorial Museum).

Aerial view of Hiroshima bombing, with details on damages to specific sites numbered 1–30. Autographed by Thomas Ferebee (U.S. Army; author's collection).

Thomas F. Farrell, deputy director of the Manhattan Project; General James Davies, 313th Bomb Wing commander; General Nathan Twining, chief of the Marianas Air Force; and Rear Admiral William Purnell.

On descending from the aircraft, Tibbets was greeted by General Spaatz, who shook his hand and pinned the Distinguished Service Cross on his flight suit.

Chapter 10. The Atomic Bombing of Hiroshima

Richard Nelson (right) with Samuel S. Kloda. Nelson was the radio operator on the *Enola Gay*. Autographed by Richard Nelson (author's collection).

After the enthusiastic greeting, all crew members were given a complete medical check-up.

A formal debriefing followed, conducted by Lieutenant Colonel Hazen Payette, the 509th intelligence officer. At the meeting they were served food, lemonade and bourbon. Present at this debriefing were General Spaatz, General Nathan Twining (representing General Curtis LeMay), General James Davies, and assorted dignitaries from the Navy who were representing Admirals Halsey and Nimitz.

August 7, 1945, was another day of celebration for the U.S. military. Each member of the air and ground crews for the three Hiroshima mission airplanes was awarded a Silver Star. Captain Parsons was promoted to the rank of commodore, U.S. Navy.

(See the following seven pages for more photographs.)

```
                                          TOP SECRET
THE MAKING OF AN EXACT COPY                                    SECTION    INITIALS   COPY NO.
OF THIS MESSAGE, AND ITS                                       CHIEF OF                  1
TRANSMISSION IN LITERAL PLAIN     INCOMING MESSAGE             STAFF
TEXT ARE AUTHORIZED SUBJECT                                    DC/S OPNS             DIST. BY.
TO NORMAL PROCEDURE FOR THE           CLASSIFICATION           DC/S P&A              CAO
SAFEGUARDING OF MILITARY                TOP SECRET             DC/S THE              6819
INFORMATION.                                                                         TYPED BY
                                                               A.G.                    emd
TELECON MSG NO.    SUBJECT                                                    DATE
   FN-06-02        BOMBS AWAY REPORT 509 SRM 13 FLOWN 6 AUGUST 1945      6 August 1945
TO:                        INFO:                       FROM:
  COMGENUSASTAF (REAR) WASH                              COMGENUSASTAF GUAM
```

PASS TO COMGENUSASTAF (REAR) WASHINGTON
FROM: COMGENAAF 20 GUAM
TO: COMGENUSASTAF (GUAM)

AIMCR 5219

   Only one Bombs Away Inflight Report received as follows:

   1 A/C bombed Hiroshima visually thru 1/10 cloud with good results. Time was 052315Z. No flak or E/A opposition.

END

TOD 0352Z

DECLASSIFIED
By: Air Force Declassification Office
27 May 2010

Following the successful bombing of Hiroshima, a coded message was sent to Tinian Island and then relayed to Guam, which sent it to Washington. This "incoming message" was declassified on May 27, 2010, by the Air Force Declassification Office (Freedom of Information Act).

Chapter 10. The Atomic Bombing of Hiroshima   117

```
                                WHITE HOUSE
    DECLASSIFIED
                                  MAP ROOM


                              6 August 1945

TOP-SECRET
        FROM: ADMIRAL EDWARDS
        TO  : ADMIRAL LEAHY (EYES ONLY)

        NR  : 334

              Following information regarding MANHATTAN
        received:
              "Hiroshima bombed visually with only 1-10th
        cover at 052315Z. There was no fighter opposition
        and no flak. Parsons reports fifteen minutes after
        drop as follows:
              'Results clear cut successful in all respects.
        Visible effects greater than any test. Conditions
        normal in airplane following delivery.'"




        RECD: 061445Z



                          TOP-SECRET
                                            COPY NO. 2
```

Top-secret report to White House. Time 06:14, to Admiral Leahy. Results of Hiroshima bombing. This report was declassified on May 3, 1973 (Harry S. Truman Library & Museum).

6/20/2020   Draft statement on the dropping of the bomb | Harry S. Truman

WHITE HOUSE

MAP ROOM

6 August 1945

FROM: THE SECRETARY OF WAR
TO : THE PRESIDENT

NR : 335

Big bomb dropped on Hiroshima 5 August at 7:15 P.M. Washington time. First reports indicate complete success which was even more conspicuous than earlier test.

STIMSON

REC'D 061500Z

Harry S. Truman Library & Museum

500 W US Hwy 24
Independence, MO 64050
816-268-8200 | 800-833-1225
Fax: 816-268-8295

Museum Hours

The museum is launching a massive renovation of the museum and its exhibitions, the first major renovation in more than 20 years and the largest since the museum opened its doors in 1957.

The museum's last day open to the public was Monday, July 22, 2019. The museum is expected to re-open in the fall of 2020.

https://www.trumanlibrary.gov/library/research-files/draft-statement-dropping-bomb?documentid=NA&pagenumber=8

Top-secret report to president. Time 06:15. From the Secretary of War's Office, Stimson sent a report to the president. This report was declassified on May 3, 1973 (Harry S. Truman Library & Museum).

Enola Gay. The Boeing B-29 Superfortress lands after the atomic bombing mission on Hiroshima, Japan. Autographed by Theodore Van Kirk, Paul Tibbets, and Tom Ferebee (U.S. Air Force; author's collection).

Enola Gay on runway following its historic mission. Autographed by Paul Tibbets and Bob Lewis (U.S. Air Force; author's collection).

The *Enola Gay* returns to its parking position after the successful atomic bomb flight over Hiroshima. Autographed by Charles McKnight, pilot of *Top Secret* (U.S. Army Air Forces; author's collection).

Paul Tibbets wearing his Distinguished Service Cross, which was presented to him upon his return from the Hiroshima mission. Autographed by Paul Tibbets (U.S. Air Force; author's collection).

Chapter 10. The Atomic Bombing of Hiroshima 121

General Carl Spaatz (commander of the U.S. Strategic Air Forces in the Pacific) attends crew interrogation of the members of the 509th Composite Group who participated in the atomic bombing of Hiroshima on August 6, 1945. Photograph taken on Tinian Island, Mariana Islands. *Present:* (1) Colonel Paul Tibbets, (2) Admiral William Parnell, (3) Captain Theodore Van Kirk, (4) General Nathan Twining, (5) Lieutenant General Barney M. Giles, (6) Lieutenant Colonel Hazen Payette, (7) Brigadier General James Davies, (8) Colonel Tom Classen, (9) Lieutenant Colonel John Porter, (10) Lieutenant Colonel James Hopkins, (11) General Carl Spaatz, (12) Admiral William Parsons, (13) Sergeant Joseph Stiborik, (14) Captain Bob Lewis, (15) Major Tom Ferebee, and (16) Technical Sergeant George Caron. Autographed by Paul Tibbets (U.S. Air Force; author's collection).

The ground and flight crew of B-29 *Enola Gay* after the atomic bombing mission on Hiroshima, Japan. Colonel Paul Tibbets is in the center under the prop, wearing a cap and his Distinguished Service Cross. Autographed by Paul Tibbets (U.S. Air Force; author's collection).

Commander Frederick L. Ashworth was the weaponeer on the B-29 *Bockscar* that dropped the "Fat Man" atomic bomb on Nagasaki, Japan, on August 9, 1945. He also selected Tinian Island as a base of operations for the 509th Composite Group. After the war, he chose the Bikini Atoll as the site for Operation Crossroads (U.S. Navy; author's collection).

Chapter 11

Truman's Address to the Nation

After the Potsdam Conference in Germany, President Truman did not fly home. For safety reasons, his security agents recommended making the five-day transatlantic journey by sea. The USS *Augusta* was chosen for the journey.

A few hours after the *Enola Gay* returned to Tinian, President Truman received the long-awaited confirmation of the historic achievement. He was noted to exclaim, "This is the greatest thing in history."

President Truman arrived in Washington, D.C., at 10:30 p.m. on August 6, 1945, and made a live radio broadcast to the American people. In fact, this speech was really directed at the Japanese government. It read:

> Sixteen hours ago an American plane dropped one bomb on Hiroshima and destroyed its usefulness to the enemy. The Japanese began the war from the air at Pearl Harbor. They have been repaid many fold. And the end is not yet. With this bomb we have now added a new and revolutionary increase in destruction to supplement the growing power of our armed forces. In their present form these bombs are now in production and even more powerful forms are in development.
>
> It is an atomic bomb. It is harnessing of the basic power of the universe. The force from which the sun draws its power has been loosed against those who brought war to the Far East....
>
> We are now prepared to destroy more rapidly and completely every productive enterprise the Japanese have in any city. We shall destroy their docks, their factories, and their communications. Let there be no mistake: We shall completely destroy Japan's power to make war.... If they do not now accept our terms they may expect a rain of ruin from the air, the like of which has never been seen on earth.

Twelve hours later, Emperor Hirohito was told that a special bomb dropped from a U.S. bomber had completely destroyed Hiroshima. However, Hirohito did not accept the U.S. demands for unconditional surrender. He mistakenly believed that the Soviets would help him negotiate a peace treaty with the United States that would protect the Imperial Dynasty.

A Japanese bulletin on Tokyo radio on the evening of August 6 said, "A few B-29s hit Hiroshima city at 8:20 a.m. August 6 and fled after dropping incendiaries and bombs. The extent of the damage is now under survey."

Four hours after the bombing, two American B-29s flew over Hiroshima to photograph the extensive damage. They reported that it was impossible to obtain

Santa Fe, New Mexico, newspaper. Autographed by Charles Albury, Fred Olivi and Theodore Van Kirk (© 2021 The Santa Fe New Mexican, Inc. Reprinted with permission. All rights reserved; author's collection).

any clear photographs, as the city was covered with a thick blanket of smoke and burning fires.

Several million propaganda leaflets, prepared by the U.S. Office of War, written in Japanese, were airdropped over major cities in Japan. The following is a translation:

Chapter 11. Truman's Address to the Nation

To the Japanese people: America asks that you take immediate heed of what we say on this leaflet. We are in possession of the most destructive explosive ever devised by man. A single one of our newly-developed atomic bombs is actually the equivalent in explosive power to what 2,000 of our giant B-29s can carry on a single mission. This awful fact is one for you to ponder and we solemnly assure you that it is grimly accurate.

We have just begun to use this weapon against your homeland. If you still have any doubt, make inquiry as to what happened to Hiroshima when just one atomic bomb fell on that city.

Before using this bomb to destroy every resource of the military by which they are prolonging this useless war, we ask that you now petition the Emperor to end the war. Our President has outlined for you the thirteen consequences of an honourable surrender. We urge that you accept these consequences and begin work of building a new, better and peace-loving Japan.

You should take these steps now to cease military resistance. Otherwise, we shall resolutely employ this bomb and all other superior weapons to promptly and forcefully end the war.

Evacuate your cities now!

Although the atomic bomb story was front-page news around the world, no photographs of the bombing were released to the public. This would have to wait.

CHAPTER 12

The Atomic Bombing of Nagasaki

The Japanese government tried to minimize the news of the complete destruction of Hiroshima. This absolute denial of reality helped convince the U.S. government that a second bomb must be dropped without delay.

When President Truman arrived at the White House on August 7, 1945, a telegram from Senator Richard Russell was waiting for a reply. Senator Russell's telegram asked Truman to "hit Japan with more atomic and conventional bombs."

On August 9, 1945, Truman replied:

> I know that Japan is a terribly cruel and uncivilized nation in warfare but I can't bring myself to believe that, because they are beasts, we should ourselves act in the same manner.
>
> For myself, I certainly regret the necessity of wiping out whole populations because of the "pigheadedness" of the leaders of a nation and, for your information. I am not going to do it until it is absolutely necessary....
>
> My object is to save as many American lives as possible but I also have a humane feeling for the women and children in Japan.

As there was no indication of a Japanese surrender, a meeting in Guam was called for on August 7. Present at the meeting were Colonel Tibbets, General Spaatz, General LeMay, Admiral Purnell, Captain Parsons, and Major John F. Moynahan. They proceeded with their orders to drop another bomb.

They agreed to use the second bomb on Thursday, August 9. Their reasoning was that Japan would believe that the United States possessed many more of these bombs and was ready to use them on all Japanese cities.

The "Fat Man" bomb left Kirtland Army Air Field, Albuquerque, New Mexico, on July 26, 1945. It was transported in a C-54 transport aircraft by the 509th Composite Group's 320th Troop Carrier Squadron. Its pilot was Gilbert Dickman. It arrived in Tinian on July 28, 1945.

"Fat Man" was 128 inches long and 60 inches in diameter, weighing 10,300 pounds. It was a plutonium bomb.

For the second atomic bombing mission, Tibbets chose Major Charles "Chuck" Sweeney. Sweeney and his aircrew would carry "Fat Man" in their strike aircraft, *Bockscar*.

Accompanying Sweeney's *Bockscar*, Captain Frederick C. Bock would fly the *Great Artiste*, as the observation/instrument plane. Major James I. Hopkins would pilot the *Big Stink* (later called *Dave's Dream*). This plane was assigned

Chapter 12. The Atomic Bombing of Nagasaki

The "Fat Man" nuclear weapon was an implosion-type bomb with a solid plutonium core. The bomb was 60 inches in diameter and 128 inches long. It weighed about 10,300 pounds and had a yield equivalent to approximately 21 kilotons of TNT. Autographed by Charles Albury, co-pilot of *Bockscar* (Los Alamos National Laboratory; author's collection).

to photograph the explosion and effects of the bomb; it would also carry scientific observers. The rendezvous point for this mission would be over Yakushima, a small island near the southern tip of Kyushu.

Kokura was the primary target for the second bombing, while Nagasaki was the secondary target. Captain George W. Marquardt, flying the *Enola Gay*, would report the weather over Kokura. Lieutenant Charles McKnight, flying *Laggin' Dragon*, would report on conditions over Nagasaki.

The *Great Artiste* carried scientific observers from Project A: Walter Goodman, Lawrence Johnston, and Jess Kupferberg. It also took along the *New York Times* reporter William (Bill) Laurence.

Walter Goodman was an electrical engineer who worked on the implosion bombs with Manhattan scientists Luis Alvarez and Harold Agnew.

During World War II, Lawrence Johnston worked at the MIT Radiation Laboratory, where he invented ground-controlled approach radar. In 1944, he went to the Manhattan Project's Los Alamos Laboratory, where he invented the exploding-bridgewire detonator. He was the only civilian to witness all three atomic explosions in 1945: the Trinity nuclear test and the atomic bombings of Hiroshima and Nagasaki.

Replica of "Fat Man." Autographed by Charles Albury (U.S. Air Force; author's collection).

Jess Kupferberg helped in the transportation and assembly of the atomic bombs.

The photo plane carried two observers from England: Dr. William G. Penney of the British Mission and Group Captain Leonard Cheshire of the Royal Air Force. Dr. Penney was a member of Project A and an expert on the effects of nuclear explosions. He was professor of applied mathematics at London University. Captain Cheshire was the former commander of the "Dambusters" squadron.

This is a list of the 12 crew members of *Bockscar* and their function.

Major Charles W. Sweeney: aircraft commander
First Lieutenant Charles Donald Albury: pilot
Second Lieutenant Frederick J. Olivi: co-pilot
Captain James Van Pelt: navigator
Captain Kermit K. Beahan: bombardier
Master Sergeant John D. Kuharek: flight engineer
Sergeant Abe Spitzer: radio operator
Staff Sergeant Edward Buckley: radio operator
Sergeant Albert Dehart: tail gunner
Staff Sergeant Ray Gallagher: assistant flight engineer/scanner
First Lieutenant Jacob Beser: radio countermeasures

Chapter 12. The Atomic Bombing of Nagasaki

Wendover, Utah. *Left to right:* **Charles W. Sweeney, pilot of** *Bockscar*; **Ray Gallagher, assistant flight engineer on** *Bockscar*; **Samuel S. Kloda, historian; Paul Tibbets, pilot of** *Enola Gay*. **Autographed by Charles Sweeney (author's collection).**

Commander Frederick Ashworth: weaponeer
Lieutenant Philip M. Barnes: assistant weaponeer

At the final briefing, at 12:30 a.m., the pilots of all six aircraft were told to fly at higher altitudes because of poor weather from a distant typhoon. Once again, Ed Doll gave Beser a small piece of rice paper containing the frequencies of the four proximity devices.

Takeoff time was set at 3:47 a.m. on August 9, 1945.

During pre-flight inspection of *Bockscar*, the flight engineer, John Kuharek, notified Sweeney that an inoperative fuel transfer pump would make it impossible to use 640 gallons of fuel carried in a reserve tank. Replacing the pump would take hours. However, moving the "Fat Man" bomb to another aircraft might take just as long and was too dangerous.

Colonel Tibbets, Admiral Purnell, General Farrell, Commodore Parsons, Commander Ashworth, and Major Sweeney, with the approval of General LeMay, decided that the mission should proceed. It would be impossible to drain the unusable fuel, so it was carried from Tinian to Japan and back again. This used up even more fuel due to extra weight.

Photograph of Jacob Beser (right) with Samuel S. Kloda. Beser was the only man on both the Hiroshima and the Nagasaki missions. Autographed by Jacob Beser (author's collection).

At 3:37 a.m., *Bockscar* rolled down the runway. Sweeney not only used up the 8,500-foot runway but also needed 200 feet of grass.

At 4:00 a.m., Frederick Ashworth, chief weaponeer, replaced the green safety plugs in "Fat Man" with red ones. "Fat Man" was now armed.

Sweeney then climbed to 17,000 feet to get above the storm clouds.

At 9:00 a.m. (Marianas Time), *Bockscar* arrived at the rendezvous point over Yakushima. Sweeney circled Yakushima until the *Great Artiste* arrived at 9:12 a.m.

Big Stink, piloted by Major James Hopkins, was flying at an altitude of 39,000 feet—9,000 feet above the agreed-upon altitude. Furthermore, Hopkins was not flying in tight circles over Yakushima, as previously agreed with Sweeney and Bock; instead, he was flying 40-mile dogleg patterns.

Sweeney had orders not to stay at the rendezvous more than 15 minutes. However, Sweeney disobeyed orders. He stayed longer and used up additional fuel.

At 9:50 a.m., Sweeney did not have any contact with *Big Stink*. Sweeney and Bock left the rendezvous point and headed for the primary target, Kokura.

During this time, the weather report from the *Enola Gay*, now piloted by Captain George W. Marquardt, sent a coded message to Sweeney: Kokura was cloudy, but visibility was good.

12/2/80

Dear Mr Kloda,

Many thanks for your note of 25 Nov. Your comments were greatly appreciated. I usually don't receive such kind remarks from Canada.

There was no tape recording or wire either for that matter. I made a special disc recording on a piece of military countermeasure gear and is was taken to Guam for transcription and use of Armed Forces Radio. I heard it played the 7th Aug 45 and never since. The original has vanished - or at least for over 35 yrs we have been unable to locate it. Whoever took it could never play it again as it was cut on a special variable speed lathe and without the original equipment is was useless except for a keepsake. I am sure that one of the Armed Forces Radio technicians put it in his duffel bag and took it home with him.

Once again, thanks for your note.

Sincerely,

Jacob Beser letter to Samuel S. Kloda. Minutes after the atomic bombing of Hiroshima, Beser, holding a special disc recorder, recorded the comments by all the crew members. The disc was taken to Guam for transcription and played only once on August 7. The original disappeared (author's collection).

Bockscar and the *Great Artiste* headed toward Kokura, which was 130 miles southwest of Hiroshima. This city was home to the Kokura Arsenal, a massive collection of war industries and numerous steelworks factories.

When they arrived at Kokura, the city was heavily overcast. Drifting smoke from fires started by a major firebombing raid by 224 B-29s on nearby Yahata the day before resulted in almost zero visibility.

Jacob Beser in front of *Enola Gay* after Hiroshima mission. Autographed by Jacob Beser (U.S. Air Force; author's collection).

In every briefing, Sweeney had been given precise instructions by Tibbets: if the weather was overcast at the primary target, go to the secondary target immediately.

Sweeney (disobeying orders) tried two bombing runs, but bombardier Beahan was unable to drop "Fat Man" visually. He did not see his aiming point, which was a large weapons factory.

Chapter 12. The Atomic Bombing of Nagasaki

Flight crew of Ship 77 for the atomic bomb mission over Nagasaki, August 9, 1945. *Left to right kneeling:* Master Sergeant John D. Kuharek, flight engineer; Staff Sergeant Ray G. Gallagher, assistant flight engineer/scanner; Staff Sergeant Albert T. Dehart, tail gunner; Sergeant Abe M. Spitzer, radio operator. *Left to right standing:* Captain Kermit K. Beahan, bombardier; Captain James F. Van Pelt, navigator; Lieutenant Charles D. Albury, airplane commander; Lieutenant Fred J. Olivi, co-pilot; Major Charles W. Sweeney, commander of 393rd Bombardment Squadron and pilot. Missing: Lieutenant Jacob Beser, radio countermeasures; Lieutenant Phillip M. Barnes, assistant weaponeer; and Staff Sergeant Edward Buckley, radio operator. Autographed by James Van Pelt, Charles Albury, Fred Olivi, and Charles Sweeney (U.S. Air Force; author's collection).

Sweeney, besides having lost precious fuel, had two other problems to contend with. First, Japanese anti-aircraft fire was getting closer. Second, Jacob Beser, who was monitoring Japanese communications, reported activity on the Japanese fighter direction radio bands.

Radar operator Sergeant Edward Buckley told Sweeney on the intercom, "Jap Zeros coming up at us. Looks like about ten."

Sweeney ignored the news of approaching Japanese Zeros and tried a third bombing run over Kokura. When Beahan did not find the target, Ashworth advised Sweeney to turn to Nagasaki. Luckily, Sweeney just escaped the Zeros.

At 11:30 a.m., Sweeney moved quickly to the secondary target. Nagasaki was 95 miles away. Sweeney was burning up fuel at the rate of 400 gallons per hour.

Photo of *Bockscar* before the Nagasaki mission. Autographed by Fred Olivi, co-pilot; James Van Pelt, navigator; Charles Albury, co-pilot; and Chuck Sweeney, pilot (U.S. Air Force; author's collection).

Photo of Samuel S. Kloda (left) with Charles Albury. Autographed by Charles Albury (author's collection).

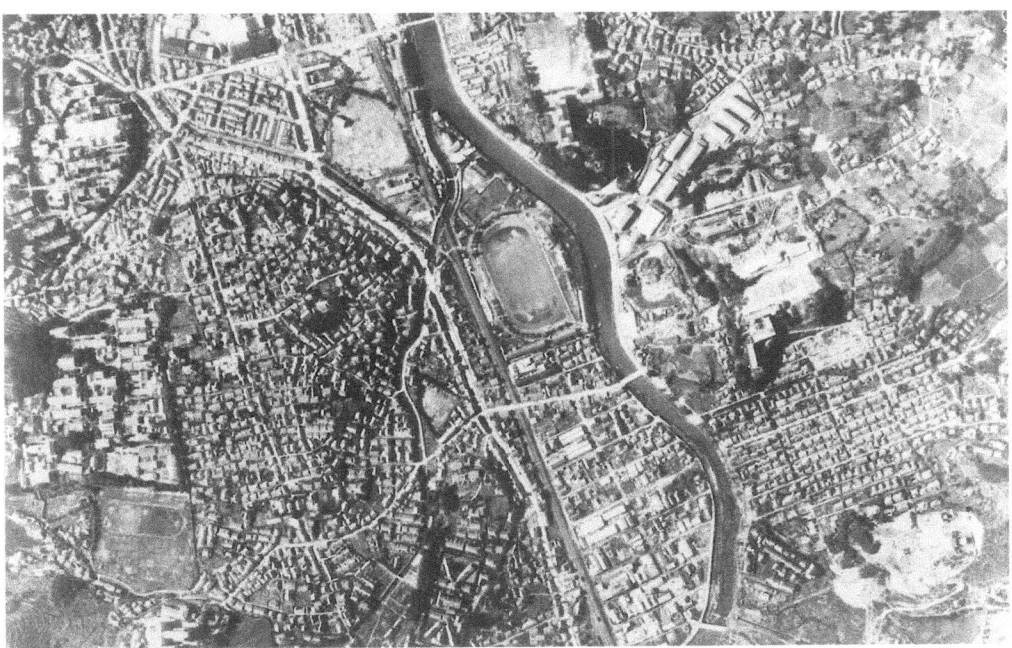

Photograph of "Ground Zero" before the atomic bombing of Nagasaki, Japan. Stadium was the aiming point. Autographed by Abe Spitzer (U.S. Air Force; author's collection).

Bockscar and the *Great Artiste* flew across the Shimonosaki Straits and followed the coastline down toward Nagasaki. During the flight to Nagasaki, Sweeney and Ashworth calculated that the remaining fuel supply was enough for one bombing run at Nagasaki and a landing at Okinawa.

When Nagasaki was finally in sight, Sweeney lowered his altitude from 32,000 feet to 28,000 feet. He started his bombing run at 200 miles per hour. The run lasted five minutes.

Here, too, there were clouds everywhere. Ashworth (who was General Groves' deputy on the flight) ordered Sweeney not to wait until the clouds cleared. He told Sweeney to make a radar approach instead of a visual attack.

However, at the last minute, Beahan found what he believed was his aiming point: the city's stadium. He released "Fat Man" at 11:02 a.m. (Nagasaki Time). In actual fact, Beahan had missed his target, causing much less damage than had originally been planned.

After falling for 43 seconds, "Fat Man" exploded at 1,650 feet above the ground. This bomb blast was far greater than that of "Little Boy." It had the equivalent power of 21 kilotons of TNT.

After Sweeney completed his 155-degree turn, shock waves hammered his aircraft. The first came from the explosion itself. The second shock wave came from the reflection under the airburst. The third shock wave came from a wall of hills. (The crew would be told later that they missed the target by one and a half miles, instead hitting the Urakami Valley.)

Captain Fred Bock, flying the *Great Artiste*, dropped instruments attached

Aerial view of Nagasaki, Japan. Photograph shows the two principal targets after the atomic bombing. Autographed by Abe Spitzer (U.S. Air Force; author's collection).

to three parachutes. These would collect information on gamma radiation, shock waves and heat.

Sweeney circled the smoke cloud once and headed toward Okinawa, with Captain Bock following.

Although the target was missed, "Fat Man" did a sizable amount of damage

Chapter 12. The Atomic Bombing of Nagasaki

to the Mitsubishi plants. In particular, it destroyed the Mitsubishi-Urakami-Ordnance works. This factory had manufactured the Type 91 torpedoes released in the attack on Pearl Harbor.

The *Bockscar*'s co-pilot, Lieutenant Colonel Fred Olivi, gave a detailed description of the explosion in his book, *Decision at Nagasaki: The Mission That Almost Failed*:

> Suddenly, the light of a thousand suns illuminated the cockpit. Even with my dark welder's goggles, I winced and shut my eyes for a couple of seconds. I guessed we were about seven miles from "ground zero" and headed directly away from the target, yet the light blinded me for an instant. I had never experienced such an intense bluish light, maybe three or four times brighter than the sun shining above us.
>
> I've never seen anything like it! Biggest explosion I've ever seen.... This plume of smoke I'm seeing is hard to explain. A great white mass of flame is seething within the white mushroom shaped cloud. It has a pinkish, salmon color. The base is black and is breaking a little way down from the mushroom. The Mushroom cloud was coming right at us. I immediately looked up and could see that he was right, the cloud was getting close to *Bockscar*. We had been told not to fly through the atomic cloud because it was extremely dangerous to the crew and aircraft. Knowing this, Sweeney put *Bockscar* into a steep dive to the right, away from the cloud, throttles wide open. For a few moments we could not tell if we were out-running the ominous cloud or if it was gaining on us, but gradually we pulled away from the dangerous radioactive cloud before it engulfed us, much to everyone's relief.

Nagasaki co-pilot Fred J. Olivi (right) with Samuel S. Kloda. Autographed by Fred Olivi (author's collection).

Dr. Tatsuichiro Akizuki was a survivor of the Nagasaki bombing. At the time, he was the medical director of the Urakami Dai-Ichi Hospital in Nagasaki. Akizuki provided medical treatment to hundreds of injured men, women and children. He wrote about the minutes following the blast in his book *Nagasaki Genbakuki* (published in English as *Nagasaki 1945*):

> All the buildings I could see were on fire.... Electricity poles were wrapped in flame like so many pieces of kindling.... It seemed as if the earth itself emitted fire and smoke, flames that writhed up and erupted from underground. The sky was dark, the ground was scarlet, and in between hung clouds of yellowish smoke. Three kinds of color—black, yellow, and scarlet—loomed ominously over the people, who ran about like so many ants seeking to escape.... It seemed like the end of the world.

Meanwhile, Major Hopkins, in *Big Stink*, made a complete circuit of the island of Kyushu. He saw a smoke cloud 100 miles away and flew to observe the explosion. He arrived in time to photograph the effects of the blast, operating a high-speed camera at an altitude of 39,000 feet. (He was supposed to film at 30,000 feet.)

About 40,000 Japanese people died immediately, and 60,000 later died of injuries. At the end of one year, the death toll from the Nagasaki bombing had reached 140,000.

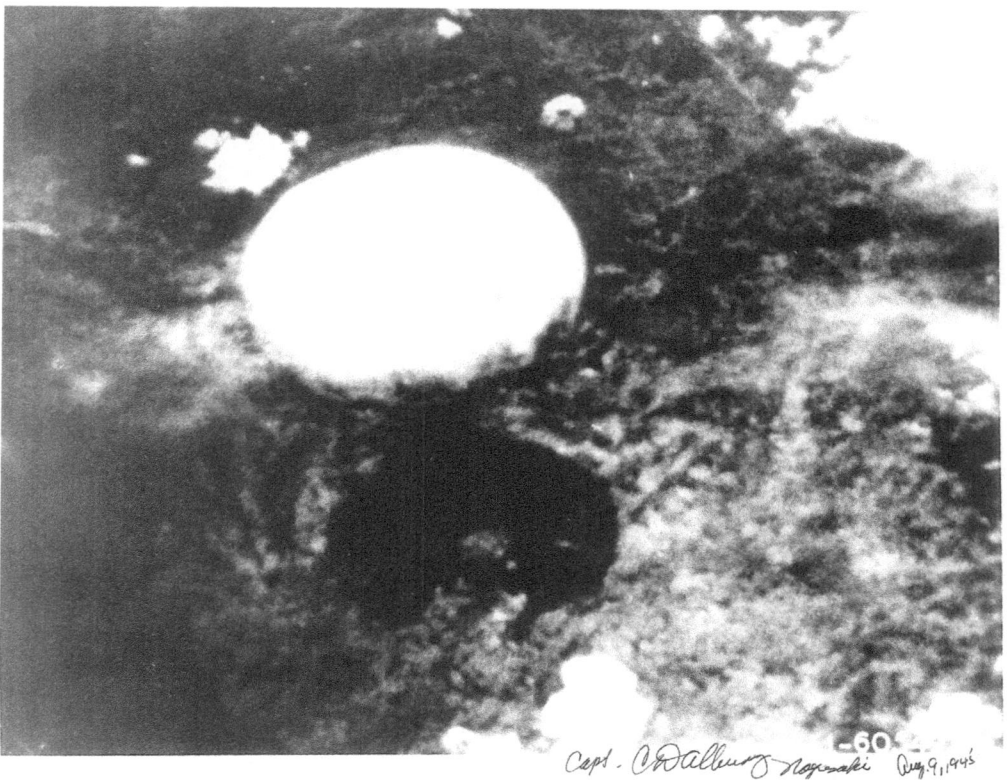

Atomic bomb burst over Nagasaki, Japan, August 9, 1945. Autographed by Charles Albury (U.S. Air Force; author's collection).

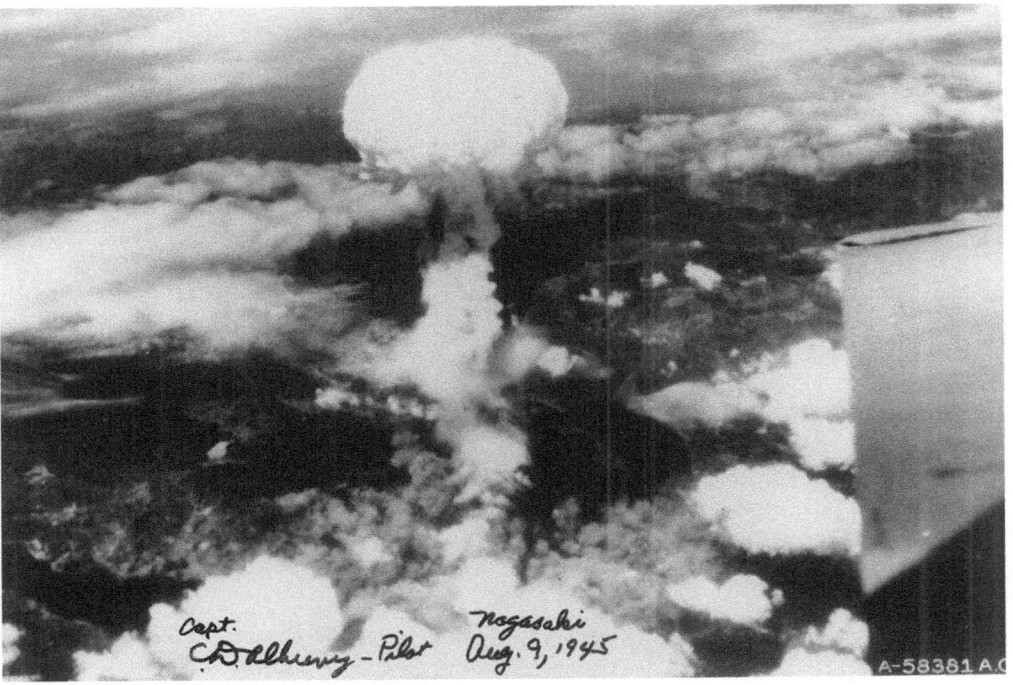

Atomic bomb mushroom cloud over Nagasaki. Autographed by Charles Albury (U.S. Air Force; author's collection).

Atomic bomb cloud rising over Nagasaki. Autographed by Frederick Ashworth (U.S. Air Force; author's collection).

Chapter 12. The Atomic Bombing of Nagasaki

This photograph shows the atomic bomb mushroom cloud over Nagasaki (Library of Congress).

As he left Nagasaki, Sweeney broke radio silence and declared, "MAYDAY, MAYDAY." Sweeney was afraid of running out of fuel and being forced to ditch the plane. All crew members donned their flotation devices in case of a forced water landing.

Opposite, top: Atomic bomb burst over Nagasaki. Autographed by Charles Sweeney (U.S. Air Force; author's collection). *Opposite, bottom:* Atomic bomb burst over Nagasaki. Autographed by Kermit Beahan (U.S. Air Force; author's collection).

Main Street, Nagasaki, Japan. A street in a residential area reduced to nothing. This was 1,000 feet northeast of the atomic bomb burst. Autographed by Charles Albury (U.S. Air Force; author's collection).

Aerial view of Nagasaki after atomic bombing (Library of Congress).

Chapter 12. The Atomic Bombing of Nagasaki 143

This photograph shows the damage to the Mitsubishi Steel and Arms Works in Nagasaki after the atomic bombing. Autographed by Kermit Beahan (U.S. Air Force; author's collection).

Sweeney was an excellent pilot. To decrease fuel consumption, he lowered his flying altitude and slowed the speed of the propellers.

As *Bockscar* approached Okinawa, Sweeney was unable to communicate with the tower at Yontan Airfield for landing instructions. He ordered emergency flares to be fired out of the plane's upper porthole. The tower saw this signal, and the runways were immediately cleared.

At 1:51 p.m., *Bockscar* landed with no fuel left in its tanks. The number two engine died from fuel starvation as Sweeney began the final approach. Touching down on only three engines midway down the landing strip, *Bockscar* bounced up into the air, about 25 feet, before slamming back down hard. The heavy B-29 skewed left, toward a row of parked B-24 bombers, before the pilots managed to regain control. Its reversible propellers were insufficient to slow the aircraft adequately, and, with both pilots standing on the brakes, *Bockscar* made a swerving 90-degree turn at the end of the runway to avoid running off it. A second engine died from fuel exhaustion before the plane came to a complete stop.

General Farrell, General Spaatz and General LeMay were notified of the successful mission.

At Okinawa, *Bockscar* was refueled and, after a five-hour flight, returned

Capt. Charles D. Albury – Pilot
Nagasaki Aug. 9, 1945

SC 244366

A general panoramic view from the Nagasaki Medical School and Hospital, looking Southeast, shows the effects of the atomic bomb explosion which took place at 7:50 A.M. on August 9, 1945. In the foreground, at the foot of the hill on which the Medical School is located, the double-track street railway loop to the hospital buildings can be seen. There was practically no damage to the tracks themselves, but the trolley wires were knocked down and the tracks covered with debris. The entire area shown in this picture was covered with industrial buildings and small residences almost as close together as it was possible to build them. In the background, the skeleton remains of the Mitsubishi Steel and Arms Works can be seen. The reinforced concrete office buildings still stand amid the wreckage of other factory buildings. In the reinforced concrete school building at the foot of the hills in the background, the doors, windows, ceiling, and building contents were damaged by the blast. Nagasaki, Japan. 6-17-46

Signal Corps Photo, released by BPR 6-17-46, Special Release 7-9-46. (Fig. 16)
Copy Neg   Lot 13950   rjh/eed

Lest we forget,
F. C. Bock August 9, 1980

Jacob Beser Radar countermeasures Officer
Aug 14, 1985 – I was in Nagasaki on
Aug 9, 1985 making a Documentary Film
for PBS. Waded for tv in Jan 1986

James F. Van Pelt Jr. Navigator –

George W Marquardt Airplane Commander
The Primary target was Kokura
When Sweeney arrived the cloud covered
cleared his landing area. Therefore
he had to bomb the secondary target
which was Nagasaki. Signed Marquardt

Capt. Charles D. Albury – Pilot
Nagasaki Aug. 9, 1945

View of Nagasaki after atomic bombing. Autographed by Charles Sweeney (U.S. Air Force; author's collection).

to Tinian. There was no homecoming party. Every crew member was totally exhausted after a 17-hour mission.

On Tinian, there was a debriefing with Admiral Purnell, who served as the Navy liaison on the Military Policy Committee for the Manhattan Project. He also represented the Navy on Tinian. Purnell told the *Bockscar* crew that they had

Opposite page: Nagasaki damage photo (released June 17, 1946). A general panoramic view from the Nagasaki Medical School and Hospital looking southeast shows the effects of the atomic bomb explosion that took place at 11:02 a.m. on August 9, 1945. In the foreground, at the foot of the hill on which the medical school was located, the double-track street railway loop to the hospital buildings can be seen. There was practically no damage to the tracks themselves, but the trolley wires were knocked down and the tracks covered with debris. The entire area shown in this picture was covered with industrial buildings and small residences almost as close together as it was possible to build them. In the background, the skeletal remains of the Mitsubishi Steel and Arms Works can be seen. The reinforced concrete office buildings still stood amid the wreckage of steel-frame buildings. In the reinforced concrete school building at the foot of the hills in the background, the doors, windows, ceiling, and building contents were damaged by the blast. Autographed on front by Charles Albury. Autographed on back by Fred Bock, Jacob Beser, James Van Pelt, George Marquardt and Charles Albury (U.S. Army; author's collection).

Chapter 12. The Atomic Bombing of Nagasaki

Nagasaki assistant flight engineer Ray Gallagher (right) with Samuel S. Kloda. Autographed by Ray Gallagher (author's collection).

missed their target, but he pointed out, "This is perhaps a most fortunate happening; for if we had hit the city, loss of life would have been far greater."

The next morning, Major Charles Sweeney received the Air Medal from General Jim Davies. Each member of the air and ground crews of the three Nagasaki mission airplanes was awarded a Silver Star.

The mission was declared a complete success.

See the following 6 pages for more photographs.

Opposite, top: Damage to Nagasaki (assessment is limited by 25 percent cloud coverage). Autographed by James Van Pelt (U.S. Air Force; author's collection). *Opposite, bottom:* Atomic bomb damage to Nagasaki. Autographed by Kermit Beahan (U.S. Air Force; author's collection).

THE MAKING OF AN EXACT COPY OF THIS MESSAGE, AND ITS TRANSMISSION IN LITERAL PLAIN TEXT ARE AUTHORIZED SUBJECT TO NORMAL PROCEDURE FOR THE SAFEGUARDING OF MILITARY INFORMATION.	**TOP SECRET** INCOMING MESSAGE CLASSIFICATION TOP SECRET	SECTION / INITIALS / COPY NO. 4 CHIEF OF STAFF DC/S OPRS — DIST. BY DC/S P&A — HS 6819 DC/S TNG — TYPED BY A.G. — vl

TELECON MSG NO.	SUBJECT		DATE
FN-09-16	Crew Observations from Reconnaissance Plane on Centerboard 509th SBM 16, Flown 9 August 1945		9 August 1945
TO: COMGENUSASTAF	INFO: COMGENUSASTAF REAR BOMWG 313 for 509th Group	FROM: COMGENAAF 20	

091541Z
AIMCR 5480

Attn: General Norstad

Approach to Nagasaki by reconn plane was from the south at 32,000 feet altitude. Column of smoke extended to an estimated altitude of 19,000 feet with black smoke covering most of the city. Approximately 20 large fires were seen through the smoke in the city area with the greatest concentrations in and around aiming poing local at 11306 on Lithomosaic, Nagasaki area, Mitsubishi Steel and Arms Works, Target No. 90.36-546. Large explosions were visible at approximately 084071 and 094079 as reconn plane passed over target at 090522Z. No apparent damage to installations on southwest side of harbor and to three large and three small vessels anchored in harbor. Clouds and smoke in the northern area of city prevented further observations. No turbulence encountered over target. Reconnaissance plane landed at northwest field, Guam, because of weather causing delay in receipt of photography. Estimate preliminary damage assessment report ready at 092100Z depending upon quality and coverage of photography.

...End...
TOD: 091606Z

DECLASSIFIED
By: Air Force Declassification Office
27 May 2010

TOP SECRET

ACTION — DC/S TNG (4 copies to Col. Fisher)

CLASSIFICATION — TOP SECRET — CLASSIFICATION

On August 9, 1945, General Lauris Norstad received confirmation of the successful Nagasaki bombing. Declassified by Air Force Declassification Office on May 27, 2010 (author's collection).

Chapter 12. The Atomic Bombing of Nagasaki

Upon the arrival at Tinian Island, Paul Tibbets (right) congratulated Charles Sweeney on his successful atomic bombing of Nagasaki. Autographed by Charles Sweeney and Paul Tibbets (U.S. Air Force; author's collection).

Mission 13: Hiroshima, Japan Atomic Bomb Attack, 6 Aug 1945.
              Twentieth Air Force

COMPOSITION OF FORCE: 1 B-29 "Enola Gay" (pilot: Col Paul W. Tibbetts);
    2 B-29 photography and observation aircraft "The Great Artiste"
    (pilot: Maj Charles W Sweeney) and "No. 91" (pilot: Capt George
    W Marquardt).

DEPARTURE: "Enola Gay" at 0245 hours, followed at 0247 and 0249 by
    other two B-29s.

PRIMARY TARGET: Hiroshima Industrial Area and City

    SECONDARY TARGET: Kokura Arsenal and City

    TERTIARY TARGET: Nagasaki Mitsubishi Steel & Arms Works and City

AIMING POINT: 063096, Litho-Mosaic Hiroshima Area

FLIGHT ROUTE:
    North Field, Tinian

    Iwo Jima (Rendezvous Point)

    Departure Point (3337N-13430E)

    Initial Point (3424N-1330530E)

    Target Area (Bomb dropped at 0915 [0815 Japan Time]; detonation
        50 seconds later).

    Breakaway (turn of 150 degrees, 3400N-13334E)

    Iwo Jima

    North Field, Tinian ("Enola Gay" landed at 1458, followed within
        the hours by the two other B-29s.

Mission 16: Nagasaki, Japan Atomic Bomb Attack, 9 Aug 1945.
              Twentieth Air Force

COMPOSITION OF FORCE: 1 B-29 "Bock's Car" (pilot: Maj Charles W Sweeney);
    2 B-29 photography and observation aircraft "The Great Artiste"
    (pilot: Capt Frederick C Bock) and "Full House" (pilot: Maj
    James I Hopkins). Hopkins lost contact with the others.

DEPARTURE: "Bock's Car" at 0347 hours, followed at intervals by the
    two other B-29s.

PRIMARY TARGET: Kokura Arsenal and City

AIMING POINT: 104082

CHECK POINT: 3243N-13233E.

Report on Mission 13 and report on Mission 16. Autographed by Jacob Beser, the only person on both atomic bomb missions (author's collection).

Chapter 12. The Atomic Bombing of Nagasaki

SECONDARY TARGET: Nagasaki Mitsubishi Steel and Arms Works and City

FLIGHT ROUTE:

    North Field, Tinian

    Iwo Jima

    Yakoshima (3020N-~~1303E~~ *13030E*) Rendezvous Point; Arrived ~~1900~~ *0910* hours, "Bock's Car" circled awaiting arrival of the other B-29s; only "The Great Artiste" showed up. After circling, departed at 0950 for Primary Target of KOKURA.

    Kokura (Arrival at Initial Point, 3343N-131380E, at 1044 and began bombing run; target obscured by smoke and clouds; Failed to sight aiming point in two further bombing runs. After c. 45 minutes proceeded on to Secondary Target, NAGASAKI.

    Check Point 3225N-13141E

    Nagasaki (Arrival at Initial Point, 3238N-13039E, at 1150; began approach by radar;) After 20 second visual bombing run dropped bomb at 1158; detonation about one minute later.

    Breakaway - Left turn of 150 degrees, 3137N-13128E, then, after circling smoke column, departed at 1205 hours for Okinawa.

    Yontan Fld, Okinawa ("Bock's Car" landed at 1351. After refueling proceeded to home base on Tinian.)

    North Field, Tinian ("Bock's Car" arrived ~~2339~~ *2245* hours.)

FLAK ON BOTH MISSIONS: ~~nil~~ *moderately heavy over Kokura*

Fred Bock
15 Feb 90

Report on Mission 16: atomic bombing of Nagasaki. Autographed by *Great Artiste* pilot Fred Bock (with corrections to initial report) (author's collection).

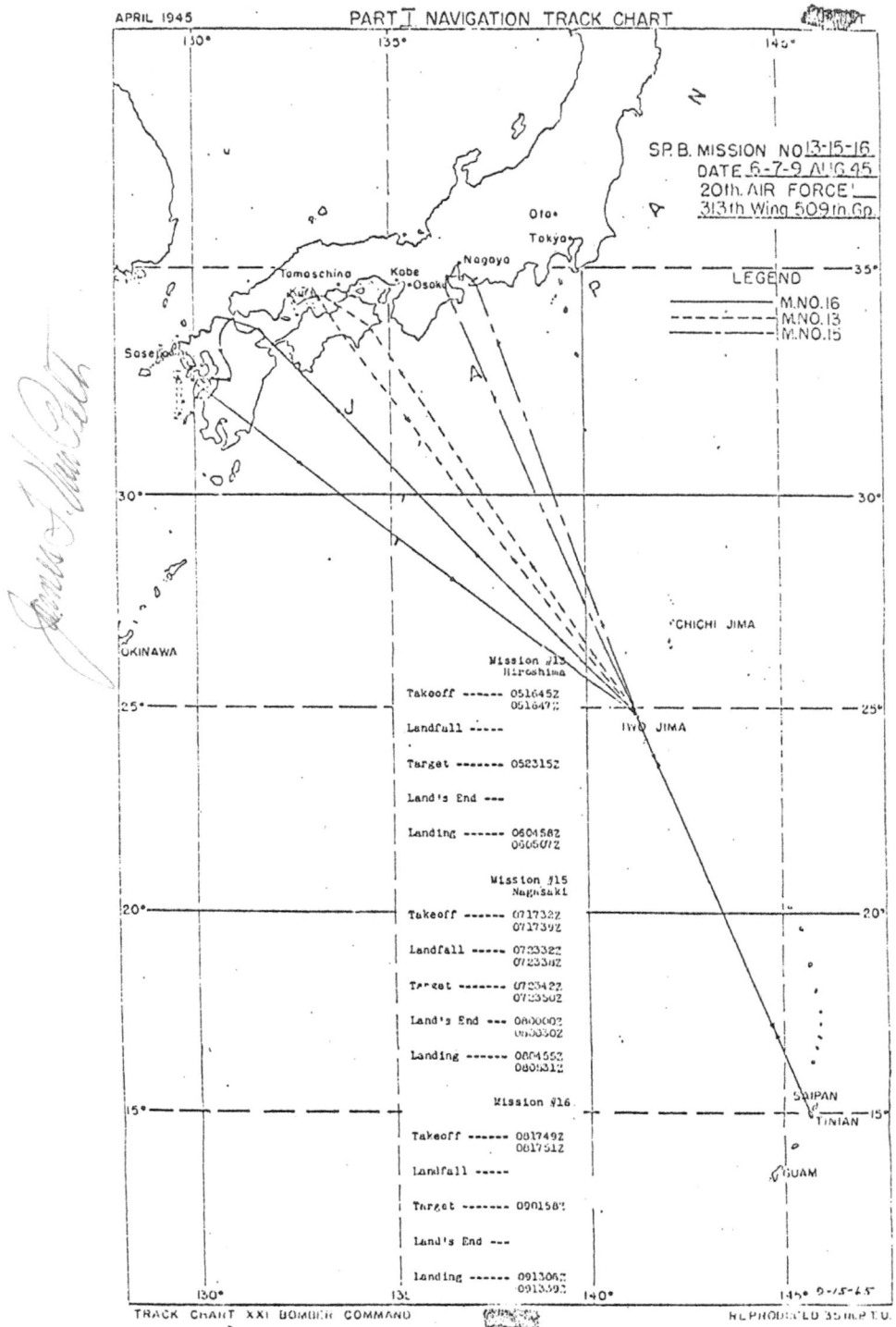

Navigation track chart. Autographed by James Van Pelt, navigator on *Bockscar* (author's collection).

Chapter 12. The Atomic Bombing of Nagasaki

General James Davies presented the Air Medal to Major Charles Sweeney (Air Force Historical Research Agency).

Chapter 13

Japan Surrenders

The Soviet Union declared war on Japan one hour before midnight on August 8, 1945. The Soviets invaded simultaneously on three fronts just after midnight on August 9. One and a half million Soviet troops crossed into the Japanese puppet state in Manchuria.

The Japanese had hoped that the Soviet Union would help them broker a deal with the Allied forces, but this was not to be.

On August 9, 1945, at 10:00 p.m., President Truman delivered a speech in the White House: "The world will note that the first atomic bomb was dropped on Hiroshima, a military base. That was because we wished in this first attack to avoid, insofar as possible, the killing of civilians. But that attack is only a warning of things to come. If Japan does not surrender, bombs will have to be dropped on her war industries and, unfortunately, thousands of civilian lives will be lost. I urge Japanese civilians to leave industrial cities immediately, and save themselves from destruction."

By August 13, Japan had not yet agreed to unconditional surrender. President Truman authorized the resumption of bombing raids on Japanese cities. The next day, August 14, 800 B-29s bombed Japan. It was called the "Big Finale."

With the crushing blow of two atomic bomb attacks and the prospect of additional cities being destroyed, Emperor Hirohito finally agreed to unconditional surrender on August 15, 1945.

Hirohito, accepting the Potsdam Declaration, gave a radio address to the people of Japan:

> To our good and loyal subjects: After pondering deeply the general trends of the world and the actual conditions obtaining in our empire today, we have decided to effect a settlement of the present situation by resorting to an extraordinary measure.
>
> We have ordered our Government to communicate to the Governments of the United States, Great Britain, China and the Soviet Union that our empire accepts the provisions of their joint declaration.
>
> To strive for the common prosperity and happiness of all nations as well as the security and well-being of our subjects is the solemn obligation which has been handed down by our imperial ancestors and which we lay close to the heart.
>
> Indeed, we declared war on America and Britain out of our sincere desire to insure Japan's self-preservation and the stabilization of East Asia, it being far from our thought either to infringe upon the sovereignty of other nations or to embark upon territorial aggrandizement.
>
> But now the war has lasted for nearly four years. Despite the best that has been done by

Chapter 13. Japan Surrenders

President Harry S. Truman and his cabinet members are seated around a conference table for a photograph on August 10, 1945. The outside border of the photograph is autographed. *From left to right:* Secretary of Agriculture Clinton Anderson, Secretary of Labor Lewis Schwellenbach, National Housing Agency Administrator John Blandford Jr., War Production Board Director Julius Krug, Office of War Mobilization and Reconversion Director John Snyder, Office of Economic Stabilization Administrator William Davis, Foreign Economic Administration Director Leo Crowley, Secretary of Commerce Henry Wallace, Under Secretary of the Interior Abe Fortas, Postmaster General Robert Hannegan, Secretary of War Henry Stimson, Secretary of State James Brynes, President Harry S. Truman, Secretary of the Treasury Fred Vinson, Attorney General Tom C. Clark, and Secretary of the Navy James Forrestal (Harry S. Truman Library & Museum).

everyone—the gallant fighting of our military and naval forces, the diligence and assiduity of out servants of the State and the devoted service of our 100,000,000 people—the war situation has developed not necessarily to Japan's advantage, while the general trends of the world have all turned against her interest.

Moreover, the enemy has begun to employ a new and most cruel bomb, the power of which to do damage is, indeed, incalculable, taking the toll of many innocent lives. Should we continue to fight, it would not only result in an ultimate collapse and obliteration of the Japanese nation, but also it would lead to the total extinction of human civilization.

Such being the case, how are we to save the millions of our subjects, nor to atone ourselves before the hallowed spirits of our imperial ancestors? This is the reason why we have ordered the acceptance of the provisions of the joint declaration of the powers.

We cannot but express the deepest sense of regret to our allied nations of East Asia, who have consistently cooperated with the Empire toward the emancipation of East Asia.

The thought of those officers and men as well as others who have fallen in the fields of

President Harry S. Truman (standing at desk) announces the surrender of Japan to reporters on August 14, 1945. *Seated on couch, left to right:* Reathel Odum; First Lady Bess W. Truman; Special Counsel Samuel Rosenman; and John W. Snyder, director of Office of War Mobilization and Reconversion. *Also pictured:* Secretary of State James Byrnes (behind President Truman, right). All others unidentified. Signed by President Truman and John Snyder (Harry S. Truman Library & Museum).

battle, those who died at their posts of duty, or those who met death [otherwise] and all their bereaved families, pains our heart night and day.

The welfare of the wounded and the war sufferers and of those who lost their homes and livelihood is the object of our profound solicitude. The hardships and sufferings to which our nation is to be subjected hereafter will be certainly great.

We are keenly aware of the inmost feelings of all of you, our subjects. However, it is according to the dictates of time and fate that we have resolved to pave the way for a grand peace for all the generations to come by enduring the [unavoidable] and suffering what is unsufferable. Having been able to save and maintain the structure of the Imperial State, we are always with you, our good and loyal subjects, relying upon your sincerity and integrity.

Beware most strictly of any outbursts of emotion that may engender needless complications, of any fraternal contention and strife that may create confusion, lead you astray and cause you to lose the confidence of the world.

Let the entire nation continue as one family from generation to generation, ever firm in its faith of the imperishableness of its divine land, and mindful of its heavy burden of responsibilities, and the long road before it. Unite your total strength to be devoted to the construction for the future. Cultivate the ways of rectitude, nobility of spirit, and work with

INSTRUMENT OF SURRENDER

We, acting by command of and in behalf of the Emperor of Japan, the Japanese Government and the Japanese Imperial General Headquarters, hereby accept the provisions set forth in the declaration issued by the heads of the Governments of the United States, China and Great Britain on 26 July 1945, at Potsdam, and subsequently adhered to by the Union of Soviet Socialist Republics, which four powers are hereafter referred to as the Allied Powers.

We hereby proclaim the unconditional surrender to the Allied Powers of the Japanese Imperial General Headquarters and of all Japanese armed forces and all armed forces under Japanese control wherever situated.

We hereby command all Japanese forces wherever situated and the Japanese people to cease hostilities forthwith, to preserve and save from damage all ships, aircraft, and military and civil property and to comply with all requirements which may be imposed by the Supreme Commander for the Allied Powers or by agencies of the Japanese Government at his direction.

We hereby command the Japanese Imperial General Headquarters to issue at once orders to the Commanders of all Japanese forces and all forces under Japanese control wherever situated to surrender unconditionally themselves and all forces under their control.

We hereby command all civil, military and naval officials to obey and enforce all proclamations, orders and directives deemed by the Supreme Commander for the Allied Powers to be proper to effectuate this surrender and issued by him or under his authority and we direct all such officials to remain at their posts and to continue to perform their non-combatant duties unless specifically relieved by him or under his authority.

We hereby undertake for the Emperor, the Japanese Government and their successors to carry out the provisions of the Potsdam Declaration in good faith, and to issue whatever orders and take whatever action may be required by the Supreme Commander for the Allied Powers or by any other designated representative of the Allied Powers for the purpose of giving effect to that Declaration.

We hereby command the Japanese Imperial Government and the Japanese Imperial General Headquarters at once to liberate all allied prisoners of war and civilian internees now under Japanese control and to provide for their protection, care, maintenance and immediate transportation to places as directed.

The authority of the Emperor and the Japanese Government to rule the state shall be subject to the Supreme Commander for the Allied Powers who will take such steps as he deems proper to effectuate these terms of surrender.

Above and following page: The Instrument of Surrender was signed on the USS *Missouri* at Tokyo Bay by Mamoru Shigemitsu (on behalf of the emperor of Japan and the Japanese government) and Yoshijirō Umezu (on behalf of Japanese Imperial General Headquarters) on September 2, 1945 (National Archives at College Park, MD).

Signed at TOKYO BAY, JAPAN at _____
on the SECOND day of SEPTEMBER, 1945.

重光葵

By Command and in behalf of the Emperor of Japan and the Japanese Government.

梅津美治郎

By Command and in behalf of the Japanese Imperial General Headquarters.

Accepted at TOKYO BAY, JAPAN at 0908
on the SECOND day of SEPTEMBER, 1945, for the United States, Republic of China, United Kingdom and the Union of Soviet Socialist Republics, and in the interests of the other United Nations at war with Japan.

Douglas MacArthur
Supreme Commander for the Allied Powers.

C.W. Nimitz
United States Representative

徐永昌
Republic of China Representative

Bruce Fraser
United Kingdom Representative

_____
Union of Soviet Socialist Republics Representative

T.A. Blamey
Commonwealth of Australia Representative

_____
Dominion of Canada Representative

_____
Provisional Government of the French Republic Representative

_____
Kingdom of the Netherlands Representative

_____
Dominion of New Zealand Representative

resolution so that you may enhance the innate glory of the Imperial State and keep pace with the progress of the world.

The formal peace treaty signing that confirmed the Japanese surrender took place on September 2, 1945. It was on board the newly built, 45,000-ton battleship USS *Missouri* in Tokyo Bay. This date would become known as V-J Day (Victory over Japan).

General Douglas MacArthur signs as Supreme Allied Commander during formal surrender ceremonies on the USS *Missouri* in Tokyo Bay. Behind General MacArthur are Lieutenant General Jonathan Wainwright and Lieutenant General Arthur E. Percival. MacArthur signed the Instrument of Surrender with six pens. Of these pens, he gave one to former POW Lieutenant General Wainwright and one to former POW Lieutenant General Percival.

The Instrument of Surrender said Japan proclaimed "unconditional surrender" to the Allied Powers, which included the United States, China, Britain and the Soviet Union (General Douglas MacArthur Memorial Archives and Library Collection).

The Japanese Instrument of Surrender was the written agreement that formalized the surrender of the Japanese Empire to the Allied Powers. The document was first signed by Japanese Foreign Minister Mamoru Shigemitsu, "By Command and on behalf of the Emperor of Japan and the Japanese Government." General Yoshijirō Umezu, chief of the Army General Staff, then signed the document, "By Command and on behalf of the Japanese Imperial General Headquarters."

U.S. General of the Army Douglas MacArthur, commander in the Southwest Pacific and supreme commander for the Allied Powers, accepted the surrender on behalf of the Allied Powers and signed in his capacity as supreme commander. After MacArthur, representatives signed the Instrument of Surrender on behalf of each of the Allied Powers: the Republic of China, the United Kingdom of Great Britain and Northern Ireland, the Union of Soviet Socialist Republics, the

A second version of the Jack Aeby color photograph of the Trinity bomb test was made and released to the public after the surrender of the Japanese, though only in black and white. Autographed by Jack Aeby (Los Alamos Photo Laboratory; author's collection).

Commonwealth of Australia, the Dominion of Canada, the Provisional Government of the French Republic, the Kingdom of the Netherlands, and the Dominion of New Zealand.

The ceremony aboard the deck of the *Missouri* lasted 23 minutes and was broadcast throughout the world.

After the ceremony was completed, the United States, for the first time, released a black-and-white photograph of the atomic bomb mushroom cloud.

Chapter 14

The Bikini Atoll Tests

At the end of World War II, Colonel Tibbets was told that the 509th Composite Group would continue, with its new headquarters in Roswell, New Mexico. Tibbets would continue to be the commander.

"Operation Crossroads" was initially intended to include three nuclear weapon tests ("Able," "Baker," and "Charlie") conducted by the United States at the Bikini Atoll in the Marshall Islands. The purpose of the operation was to investigate the effect of nuclear weapons on naval warships. Operation Crossroads was a joint Army-Navy operation with Vice Admiral William P. Brandy of Joint Task Force One as commander. General Sinclair Street would select the Air Force personnel who would participate.

In January 1946, General LeMay requested that Colonel Tibbets report to General Street at Headquarters Command at Washington's Bolling Air Force Base. Tibbets brought with him his bombardier and friend, Tom Ferebee.

At this meeting, Tibbets and Ferebee were asked several questions dealing with the altitude of the B-29 plane and potential bomb accuracy. Tibbets asked to be the pilot of the atmospheric nuclear test, with Ferebee as the bombardier. Tibbets told General Street that he and Ferebee could drop the bomb within 300 feet of any target.

Tibbets suggested that obsolete naval vessels should be placed in concentric circles around the target. This arrangement would give evaluators a good indication of how destructive the bomb could be against ships. Each ship would be fitted with equipment to measure the radiation from the bomb.

Unfortunately, because of infighting and extreme jealousy, Tibbets was relieved of command of the 509th Composite Group. Tibbets accepted his demotion, but he still wanted to be the Crossroads pilot. The Air Force then decided to have a "Bomb-Drop Competition." Whoever came closest to the target would win.

The competition took place in Albuquerque, New Mexico. Tibbets flew a B-29. From an altitude of 30,000 feet, Ferebee averaged a 237-foot error after dropping 14 inert bombs. A second crew averaged an error of over 380 feet, while a third crew averaged an error of over 700 feet.

It was obvious to everyone that Tibbets and Ferebee had won the competition. However, as Tibbets recalled, "A couple of operations analysts were called in and they figured—making allowance for an intangible and completely irrelevant

"Operation Crossroads" test, Roswell Army Air Field, New Mexico. Brigadier General Roger M. Ramey (seated) was commander of the Army Air Force units that participated in the atomic bomb experiments upon expendable fleet units at Bikini Atoll, Marshall Islands. *Left to right:* Colonel Alfred F. Kalberer, in charge of intelligence; Colonel William Blanchard, in charge of the bombardment; and Colonel Paul W. Tibbets, technical director of the project. Autographed by Paul Tibbets (U.S. Air Force; author's collection).

factor called 'ballistic winds'—that if the bombs had been dropped over the Pacific, the third crew would have come closest. In fact, they figured Tom and me right into last place."

Instead of choosing Tibbets and Ferebee, the competition winners were Major Woodrow P. Swancutt (pilot), Major Harold H. Wood (bombardier), Captain William C. Harrison (co-pilot), and Major William B. Adams (navigator).

All tests used the "Fat Man" plutonium implosion-type nuclear weapon. Each bomb had an equivalent yield of 23 kilotons of TNT.

These tests required many men and equipment: a fleet of 242 ships, 42,000 men, 156 airplanes, and tens of thousands of ordnance pieces. Unfortunately, as a result of the tests, 167 natives of Malayo-Polynesian origin who made a living by growing native crops and diving for shellfish were moved from the Bikini Atoll to the Rongerik Atoll, some 125 miles eastward.

Forty-two thousand guests from around the world (along with 124 writers) were invited to the first test, scheduled for July 1, 1946.

During the Operation Crossroads test, Theodore Van Kirk, Paul Tibbets, and Tom Ferebee completed a practice mission of dropping a dummy bomb within 25 feet of the target. Autographed by Paul Tibbets (U.S. Air Force; author's collection).

The purpose of the tests was clearly explained by the commander, Vice Admiral Brandy, in "Operations Crossroads Background Material": "The tests stand out clearly as a defensive measure. We are seeking to primarily learn what types of ships, tactical formations and strategic

Major Woodrow P. Swancutt was a Boeing B-29 Superfortress pilot in the 40th Bombardment Group, which was the first B-29 unit to fly in battle. After the war, Swancutt was selected as pilot on *Dave's Dream* to drop the nuclear weapon during the "Able" test in Operation Crossroads. Later in his career he was promoted to major general. Autographed by Woodrow P. Swancutt (U.S. Air Force; author's collection).

At Roswell Army Air Field in New Mexico, Army Air Force technicians check control equipment for radio-controlled Boeing B-17 Flying Fortress test hop as the AAF continues preparing for its participation in the joint Army-Navy atomic bomb test. The B-17 would fly into the atomic bomb cloud to gather scientific data. Autographed by William C. Harrison, co-pilot of *Dave's Dream* (U.S. Air Force; author's collection).

dispositions of our own naval forces will best survive attack by the atomic weapons of other nations, should we ever have to face them." The tests would assess the effects of pressure, impulse, shock-wave velocity, optical radiation, and nuclear radiation of a plutonium bomb.

The first test, called "Able," was conducted on July 1, 1946, at 9:00 a.m. For this test, the USS *Nevada* was the designated target ship, while 24 vessels were placed within a 1,000-yard radius. The USS *Nevada* was selected chiefly because it was the most rugged battleship available.

Seventy-eight vessels would be in the target zone for the first test. The vessels included U.S. battleships, 2 aircraft carriers, 2 cruisers, 13 destroyers, 8 submarines, numerous auxiliary and amphibious vessels, 3 surrendered German ships, and 48 Japanese warships. These 48 warships of the Imperial Japanese Navy were seized by the United States after the end of World War II.

In a Senate hearing on August 25, 1945, Senator Brien McMahon had been questioned on the new weapon's effect on ships. He stated, "In order to test the destructive powers of the atomic bomb against naval vessels, I would like ...

"Crossroads Project." Radio transmitters installed in Boeing B-29 for broadcast description of atomic bomb tests at Bikini Atoll. Photograph taken at March Field, California, May 6, 1946. Autographed by Woodrow P. Swancutt, pilot of *Dave's Dream* (U.S. Air Force; author's collection).

Japanese naval ships taken to sea and an atomic bomb dropped on them. The resulting explosion should prove to us just how effective the atomic bomb is when used against the giant naval ships. I can think of no better use for these Jap ships."

For the "Able" test, the bomb (named "Gilda") was dropped from a B-29 Superfortress called *Dave's Dream* of the 509th Bombardment Group. (In April

Chapter 14. The Bikini Atoll Tests

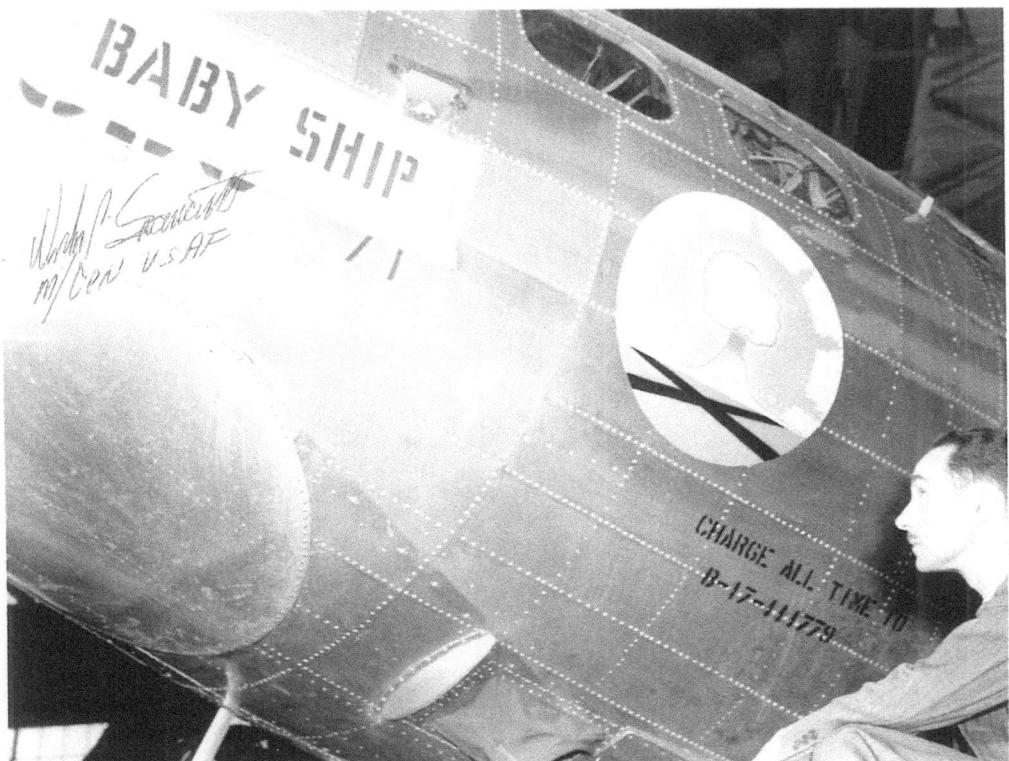

Remote-control aircraft. Mr. Holland of the Wright Field Motion Picture Laboratory views an insignia of the Army Air Forces "Crossroads Project," which has been painted on one of the baby drones. This plane was one of the 10 modified Boeing B-17s that was flown by remote control during the atomic bomb tests at Bikini Atoll. Autographed by Woodrow P. Swancutt, pilot of *Dave's Dream* (U.S. Air Force; author's collection).

1946, the *Big Stink* aircraft was assigned to Operation Crossroads and renamed *Dave's Dream* by its crew in honor of Captain David Semple, who had been killed in the crash of another B-29 on March 7, 1946, near Albuquerque, New Mexico. Semple had been a bombardier in many of the 155 test drops for the Manhattan Project.)

The USS *Nevada* was painted a bright orange for clear visibility from 30,000 feet. The proximity fuse detonated the bomb 520 feet above the target fleet and caused less than the expected amount of damage because it missed the target by 2,130 feet.

By missing its target, the bomb blew out one of the main instrumentation ships. Only two vessels sank; one was capsized, and 18 were damaged. Most surviving target ships were reboarded within 24 hours.

The government called for an investigation of the flight crew, wanting to know why the target ship was not hit. (Tibbets believed it was pilot error.) The effects of radiation on the vessels were almost totally ignored.

The Army/Navy called "Able" a complete success.

The second test, "Baker," occurred on July 25, 1946, at 8:35 a.m. The test

B-29 final engine check. The 58th Wing boasted the highest maintenance efficiency record in the Army Air Forces. Autographed by William C. Harrison, co-pilot of *Dave's Dream* (U.S. Air Force; author's collection).

bomb (named "Helen of Bikini") was encased in a watertight steel caisson and suspended 90 feet beneath the landing ship *LSM-60*. Seventy-five obsolete vessels were placed in the target zone.

In order to gather radiation samples, eight B-17s were converted into remote-controlled drones. Their pilots operated from a mother plane at a safe

Army Air Forces demonstrated a radio-controlled Boeing B-17 Flying Fortress. The aircraft was used in Operation Crossroads to gather scientific data. Autographed by Woodrow P. Swancutt, pilot of *Dave's Dream* (U.S. Air Force; author's collection).

distance from the detonations. The drones flew directly into the mushroom cloud to take pictures and collect the samples.

The early morning bomb detonation caused a blast to displace 2.2 million cubic yards of aquatic material and created a 30-foot-deep crater with a maximum diameter of 1,100 yards; the segment of the crater deeper than 20 feet covered an area 250–700 yards in diameter.

Beginning four milliseconds after the explosion, the shock wave reached the surface, throwing up a "spray dome," rising at an initial speed of 2,500 feet per second (Mach 2.5), in advance of the rapidly expanding bubble of hot gases. This caused a tsunami that generated a wave 94 feet high.

Ten seconds after detonation, water started to escape the stem and fall back toward the surface. At its greatest extent, the column was 2,000 feet across, with walls 300 feet thick and 6,000 feet tall, holding a million tons of water.

Within four minutes, the mushroom cloud head was 6,000 feet across. It rose to 10,000 feet before dispersing. The base surge was prominent. Expanding outward initially at over 60 miles per hour, it rapidly rose. The expanding surge formed a doughnut-shaped ring 3.5 miles across and 1,800 feet high.

The detonation of "Baker" created such pressure under water that huge

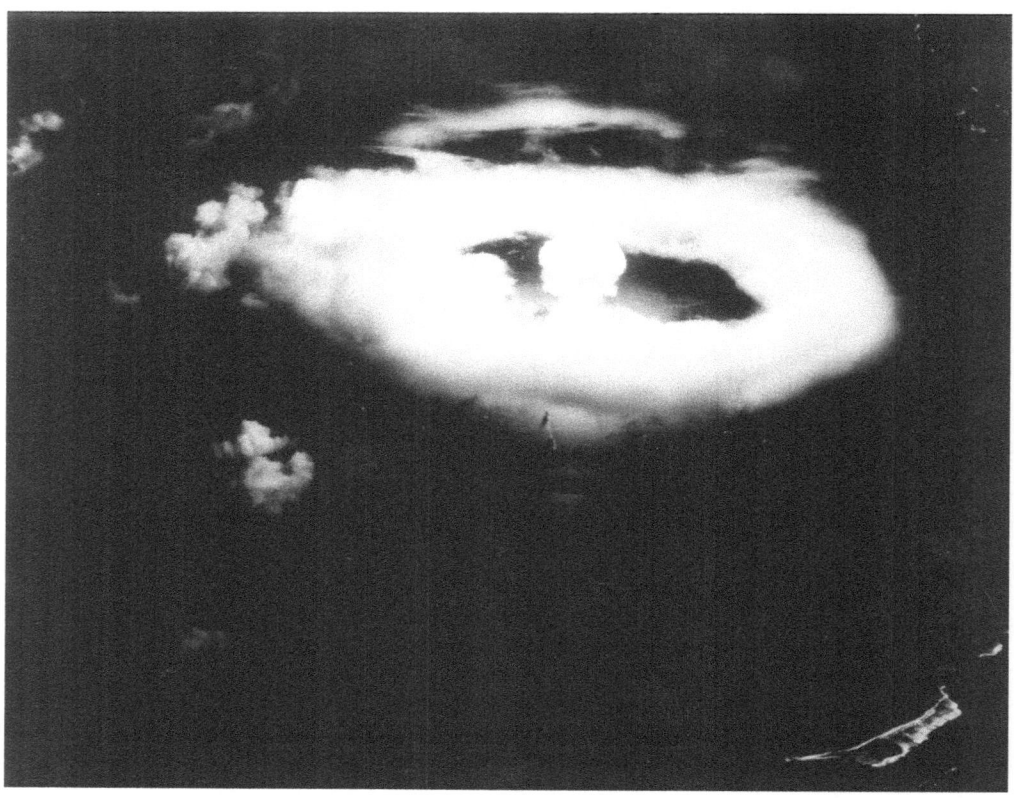

First atomic bomb explosion at Bikini Atoll in the Marshall Islands, July 1, 1946 ("Able" test). This was the fourth atomic bomb to be exploded. Autographed by Woodrow P. Swancutt, pilot of *Dave's Dream* (U.S. Air Force; author's collection).

splashes from the bottom of the lagoon sprayed radioactive water over most of the 75 vessels.

When this water fell back into the lagoon, it sank nine vessels immediately.

The "guests" and Navy were all impressed with the bomb blast and the destruction of the various ships. However, far deadlier than the actual blast was the lasting effect of radiation. All of the ships that were "undamaged" would suffer from radioactive contamination. Radioactive material adhered to all parts of the ships. Every means of removing or even reducing the radiation levels was fruitless.

The Navy realized that the situation was serious. The "severe" contamination problem was kept "top secret."

On September 4, 1946, the Commander in Chief Pacific Fleet (CINCPAC) ordered the sinking of 83 vessels at Kwajalein. Only nine vessels had "survived."

The Army/Navy called "Baker" a complete success. However, Glenn Seaborg ("The Father of Plutonium"), the longest-serving chairman of the Atomic Energy Commission, called the "Baker" test "the world's first nuclear disaster."

The Navy now understood that a fleet of ships could physically survive a

"Able" nuclear weapons test on Bikini Atoll, showing ships. Detonation occurred at 09:01:01 a.m. The time-lapse photograph shows 09:01:39. The mushroom cloud expanded in 38 seconds (Library of Congress).

"Baker" test, July 25, 1946, Bikini Lagoon. Sweeping out in all directions, this haze of water, steam spray and radioactive substances obscured all but extreme outer fringes of the Crossroads target fleet a few minutes after the "Baker" blast. This photograph was taken by an automatic camera mounted in one of the photographic towers on a nearby island. This was the fifth atomic bomb to be exploded. Photo by joint Army-Navy Task Force One. Autographed by Woodrow P. Swancutt (U.S. Army; author's collection).

nuclear blast but could be completely lost to radioactive contamination. A chilling report by the Joint Chiefs of Staff Evaluation Board for Operation Crossroads (dated June 30, 1947) stated,

> From a military viewpoint, the atomic bomb's ability to kill human beings or to impair, through injury, their ability to make war is of paramount importance. Thus the overall result of a bomb's explosion upon the crew ... is of greater interest.... If used in numbers, atomic bombs not only can nullify any nation's military effort, but can demolish its social and economic structure and prevent their re-establishment for long periods of time. With such weapons, especially if employed in conjunction with other weapons of mass destruction, as, for example, pathogenic bacteria, it is quite possible to depopulate vast areas of the earth's surface, leaving only vestigial remnants of man's material works.

The *Washington Post* published a column by Drew Pearson that called the results of the Bikini tests a "major naval disaster."

Operation Crossroads was officially terminated on August 10, 1946, due to radiation safety concerns. Army Colonel Dr. Stafford Warren, chief of the Medical Section during the Manhattan Project, concluded that the effort to

Chapter 14. The Bikini Atoll Tests

"Baker" test: aerial view of mushroom cloud (Library of Congress).

decontaminate the target vessels for a scheduled third test, "Charlie," was impossible and dangerous.

See the following five pages for more photographs.

"Baker" test: photograph taken from Bikini Atoll. Photo by joint Army-Navy Task Force One. Autographed by David M. Critchlow (U.S. Army; author's collection).

Opposite, bottom: "Baker" test: Northeast Lagoon, Bikini Atoll. The wider, exterior cloud is a condensation cloud caused by the Wilson chamber effect. The actual mushroom cloud is inside the condensation cloud (Wikimedia Commons).

"Baker" test: cloud formation at peak shortly before the rapid disintegration of the mushroom shape and as the column of water starts to fall. Photo by joint Army-Navy Task Force One. Autographed by Woodrow P. Swancutt (U.S. Army; author's collection).

"Baker" test: this aerial photograph shows tons of water, steam and radioactive materials boiling up out of the lagoon. Photo by joint Army-Navy Task Force One. Autographed by Woodrow P. Swancutt (U.S. Army; author's collection).

Opposite, top: "Baker" test: the column of water as it began to fall. Photograph taken by an automatically operated camera on a nearby island. Autographed by William C. Harrison (U.S. Army; author's collection). *Opposite, bottom:* "Baker" test. Autographed by David Critchlow (U.S. Army; author's collection).

Chapter 14. The Bikini Atoll Tests

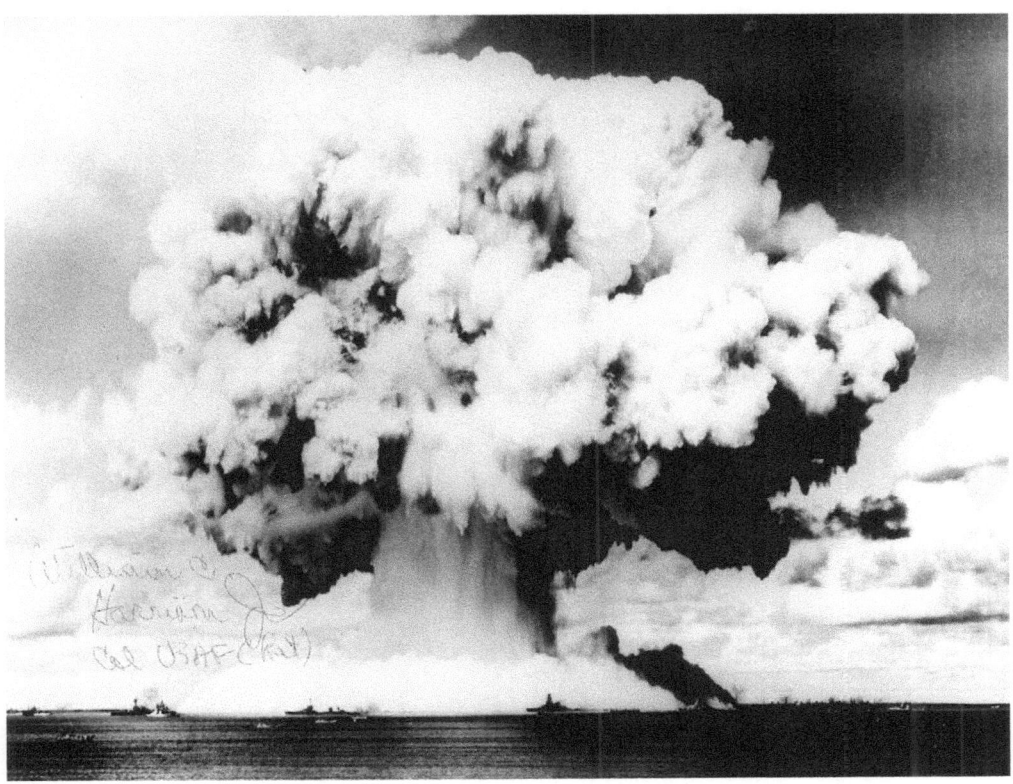

Tons of water thrown up out of Bikini Lagoon by the "Baker" blast shower down on the Crossroads target fleet. The spreading circle of radioactive materials is shown sweeping out in an ever-widening ring over the target ships. This photograph was taken by an automatic camera mounted in one of the Bikini towers. Autographed by William C. Harrison (Joint Army-Navy Task Force One; author's collection).

Epilogue: Japan

Japan

On August 15, 1945, the Japanese government accepted the Potsdam Declaration; on September 2, its representatives formally signed the Japanese Instrument of Surrender.

On September 2, 1945, President Harry S. Truman appointed General Douglas MacArthur supreme commander for the Allied Powers to supervise the occupation of Japan.

On September 6, the United States issued a document titled "U.S. Initial Post-Surrender Policy for Japan," which had been drafted by the State-War-Navy Coordinating Committee and approved by that committee on August 29. Its official designation was SWNCC 150/4, and it was approved by the president on September 6.

This document sets the two main objectives for occupation:

> The ultimate objectives of the United States in regard to Japan, to which policies in the initial period must conform, are:
> 1. To insure that Japan will not again become a menace to the United States or to the peace and security of the world.
> 2. To bring about the eventual establishment of a peaceful and responsible government which will respect the rights of other states and will support the objectives of the United States as reflected in the ideals and principles of the Charter of the United Nations. The United States desires that this government should conform as closely as may be to principles of democratic self-government but it is not the responsibility of the Allied Powers to impose upon Japan any form of government not supported by the freely expressed will of the people.

This document became the official legal document for the conduct of Japanese affairs during the occupation.

The Japanese occupation (code named "Operation Blacklist") would be completely implemented by General MacArthur. Under his authority and guidance, the Japanese government introduced extensive social and economic reforms. The Japanese constitution was completely changed to allow democracy.

Most of the Allied nations that entered World War II wanted Emperor Hirohito tried as a war criminal. However, General MacArthur believed that such a prosecution would be overwhelmingly unpopular with the Japanese population.

General MacArthur and Emperor Hirohito at their first meeting, at the U.S. Embassy, Tokyo, September 27, 1945. Autographed by both (photograph by Lieutenant Gaetano Faillace, General Douglas MacArthur Memorial Archives and Library Collection).

On September 27, 1945, Emperor Hirohito requested a visit with General Douglas MacArthur. The meeting lasted for only a few minutes. Through his personal translator, Hirohito accepted responsibility for the conduct of the war. At the end of this brief meeting, two photos of General MacArthur and Emperor Hirohito were taken by U.S. Army photographer Lieutenant Gaetano Faillace.

In the months and years that followed, Hirohito was the key to a smooth transition from a militaristic autocratic government to a Western-style democracy. Hirohito continued to visit with General MacArthur semiannually until MacArthur's retirement.

At the beginning of January 1946, there were over 350,000 U.S. personal stationed in Japan. Their first concern was feeding millions of Japanese. With the collapse of the government, virtually everyone was starving. The United States set up distribution centers and aided the farmers, both through financial grants and by purchasing farm machinery.

Operation Blacklist ended on September 8, 1951, with the signing of the San Francisco Peace Treaty, which allowed Japan to have complete sovereignty. The San Francisco Peace Treaty, signed by 48 nations, came into effect on April 28,

Chapter 14. Epilogue: Japan

Photograph taken of Hiroshima on August 6, 1946 (one year after the atomic bombing). The T-shaped bridge (Aioi), which was the aiming point for the bomb, was still standing in a rather dilapidated condition. The cement railings were blown off, and the surface of the bridge was covered with huge wrinkles and holes. In the foreground could be seen the new homes that had been built recently. Autographed by Joseph Stiborik, radio operator (U.S. Air Force; author's collection).

1952. This treaty was a bilateral decision that eventually helped secure the enduring relationship between the United States and Japan.

The treaty included the termination of the Imperial Japanese Empire. It detailed territorial as well as postwar mandates and conditions Japan had to follow in order to turn the page on the nation's gloomy past. In effect, it created a form of international rules, not through conflict and terror, but rather through peaceful dispute and deliberations.

The United States occupied Japan for over six years and helped rebuild the nation through its assistance and instilling in the Japanese the need for democracy.

Japan is now one of the world's most successful democracies and largest economies. The U.S.-Japan alliance is the cornerstone of U.S. security interests in Asia and is fundamental to regional stability and prosperity. The alliance is based on shared vital interests and values, including the maintenance of stability in the Indo-Pacific region, the preservation and promotion of political and economic freedoms, support for human rights and democratic institutions, and the expansion of prosperity for the people of both countries and the international community as a whole. Japan provides bases as well as financial and material support to U.S. forward-deployed forces, which are essential for maintaining stability in the region.

Because of the two countries' combined economic and diplomatic impact on the world, the U.S.-Japan relationship has become global in scope. The United States and Japan cooperate on a broad range of global issues, including development assistance, global health, environmental and resource protection, and women's empowerment.

The greatest display of reconciliation occurred in May 2016, when President Barack Obama became the first U.S. president to visit Hiroshima. Japanese Prime Minister Shinzo Abe reciprocated seven months later by visiting Pearl Harbor.

Hiroshima

In 1958, the population of Hiroshima reached 410,000. Today, Hiroshima is a prosperous manufacturing center with a population of over 1.2 million. The people of Hiroshima have not forgotten the fact that their city was once transformed into ash by an atomic bomb. The Atomic Bomb Dome, or Genbaku Dome, is a UNESCO World Heritage Site and stands as a silent witness to the horror of that bombing.

Every year on August 6, the anniversary of the day when "Little Boy" was dropped, Hiroshima holds a ceremony in Peace Memorial Park. Hiroshima is determined to fulfill its mission as an international peace culture city. It will continue doing everything in its power to rid our world of nuclear weapons and build a world of genuine and lasting peace.

Opposite, bottom: **Atomic Bomb Dome. Inscription by Morris Jeppson, weapons test officer on *Enola Gay* mission (collection of Clay Kemper Perkins).**

Chapter 14. Epilogue: Japan

The gutted Hiroshima Prefectural Industrial Promotion Hall, currently known as the Atomic Bomb Dome. Hiroshima Peace Memorial Museum. Autographed by Jacob Beser (U.S. Air Force; author's collection).

The alternative to the Hiroshima bombing was to continue to send hundreds of B29 bombers daily to destroy Japanese cities and military facilities — massive burning and leveling. This was WAR. And there was scheduled an immense Allied invasion of Japanese islands in November. The hope and expectation was the atomic bombing would terminate the conflict of WWII and further destruction and loss of lives.

Morris Jeppson
Weapons Test Officer
Enola Gay Mission
HIROSHIMA 6 Aug, 1945

The survivors of the Hiroshima and Nagasaki bombings call themselves *hibakusha*. In 2017, the International Campaign to Abolish Nuclear Weapons was awarded the Nobel Peace Prize. On December 11, 2017, Setsuko Thurlow, a *hibakusha*, spoke on behalf of the organization (her speech appeared in *The Mainichi*, a Japanese newspaper). Thurlow told the audience of her painful experiences in the bombing of Hiroshima 72 years earlier. She was among the population of students mobilized to work during the war and was at her workplace a little more than a mile from the hypocenter when the atomic bomb exploded over the city on August 6, 1945. She became trapped under smoldering rubble, but she heard a man call to her, "Don't give up! Keep pushing! I am trying to free you. See the light coming through that opening? Crawl towards it as quickly as you can." As she crawled out, she said, the ruins were on fire, and most of her classmates in the building were burned to death.

Thurlow also recalled her four-year-old nephew, who she said "kept begging for water in a faint voice until his death released him from agony." She said his body "had been transformed into an unrecognizable melted chunk of flesh" and added, "To me, he came to represent all the innocent children of the world, threatened as they are at this very moment by nuclear weapons. Every second of every day, nuclear weapons endanger everyone we love and everything we hold dear. We must not tolerate this insanity any longer," she said, eliciting lasting applause.

Following the war, Chuck Sweeney gave lecture tours throughout the United States, raising money for an orphanage in Hiroshima. Bob Lewis likewise helped raise funds for the medical treatment of Japanese girls who were disfigured by the intense heat of the bomb.

Nagasaki

The city of Nagasaki replaced its industrial war factories with manufacturing centers that could bring in needed currency through foreign trade. Nagasaki also expanded its seaport, shipbuilding and fishing fleet.

"Fat Man" exploded 1,500 feet above Matsuyama in Nagasaki. Today a memorial monolith stands in Hypocenter Park, which marks ground zero of the atomic explosion. Hundreds of cherry blossom trees fill the park.

Near the area are the important sites that have been preserved for future generations: the Atomic Bomb Museum, the National Peace Memorial Hall for the Atomic Bomb Victims, and Peace Park. One of the main attractions is the Urakami Cathedral, which has been restored with its atomic-bombed statue of the Virgin Mary.

Atomic Bomb #3

One of the most important secrets of World War II was that the atomic bombs destined for Japan had their final assembly at Wendover, Utah. This information was only made public in declassified documents in 1980.

Chapter 14. Epilogue: Japan

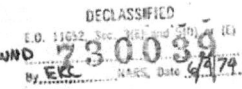

10 August 1945.

MEMORANDUM TO: Chief of Staff.

The next bomb of the implosion type had been scheduled to be ready for delivery on the target on the first good weather after 24 August 1945. We have gained 4 days in manufacture and expect to ship from New Mexico on 12 or 13 August the final components. Providing there are no unforeseen difficulties in manufacture, in transportation to the theatre or after arrival in the theatre, the bomb should be ready for delivery on the first suitable weather after 17 or 18 August.

L. R. GROVES,
Major General, USA.

General Leslie Groves sends a top-secret memorandum to General George Marshall, Chief of Staff, informing him of atomic bomb #3 (countersigned by General Marshall). Declassified on June 4, 1974 (U.S. Army).

On August 10, 1945, one day after the atomic bombing of Nagasaki, Major General Leslie Groves sent a top-secret memorandum to the Army chief of staff, General George Marshall. The single paragraph read, "The next bomb of the implosion type had been scheduled to be ready for delivery on the target on the first good weather after 24 August 1945. We have gained 4 days in manufacture

Livermore, CA
June 3, 1991

Dear Mr. Kloda,

Thank you for the address of Gilbert Dickman.

About A Bomb #3. In the morning of the day of the Japanese surrender, at 9 AM the plane carrying the bomb touched down at Wendover. It was unloaded and delivered to the assembly bldg. The assembly crew removed the nuclear core replacing it with a steel core and reassembled the bomb. It then went into an Igloo.

When we left Wendover the bomb went with us. With no place to store it on Oxnard (later Sandia) I took over the Igloos at Kirtland Field. Here it stayed for over two years.

Those few who had known about the third Bomb had returned to their former jobs or been transfered to other locations. The 3rd bomb had become the forgotten bomb. In 1947, Warner who had been the last head of the Manhatten District preceding me had come back to take over the assembley team for the A.E.C. in Operation Crossroads.

He wanted me to assemble a bomb that
(over)

Above and opposite: **June 3, 1991, letter from James Lee Rowe (head of W-47) to Samuel S. Kloda, discussing atomic bomb #3 (author's collection).**

and expect to ship from New Mexico on 12 or 13 August the final components. Providing there are no unforeseen difficulties in manufacture, in transportation to the theatre or after arrival in the theatre, the bomb should be ready for delivery on the first suitable weather after 17 or 18 August."

Because there was silence from Japan following the first and second atomic

Chapter 14. Epilogue: Japan

> could be used as a dress rehearsal prior to ~~giver going~~ Going near Bikini Atoll for their operation. This presented an opportunity to get rid of bomb #3. So the bomb, less the nuclear core was flown over the ocean west of Los Angeles (about 500 miles) and dropped. It worked perfectly.
>
> The new assembly crew took two of my men and with a new AEC Assembly crew, now. I dissbanded my explosive assembly crew. At the same time I abanded this assembly building at Kirtland Field.
>
> An unfortunate incident befell this new assembly crew.
>
> When the nuclear air drop occured the wind shifted and blanketed the ships, on which the assembly was made, with nuclear fallout.
>
> The air drop west of Los Angeles ended the Sega of the third bomb.
>
> I hope this is what you wanted to know.
>
> Sincerely
>
> James Les Rowe

bombings, Colonel Paul Tibbets was ordered by General Curtis LeMay to fly from Tinian to Wendover to pick up bomb #3. By the time Tibbets and his crew arrived at the debarkation point in California, however, the war was over.

The third atomic bomb was still at Wendover, under the control of James Les Rowe (head of Project W-47). When the war ended, Rowe and the members of the assembly crew first brought #3 to Oxnard (later Sandia) and finally to Kirtland Air Field. According to Rowe's letter of June 3, 1991, "Here it stayed for over two years."

Roger S. Warner, Jr., was a civilian scientist who worked as part of Project Albert, the team that transported and assembled all the components of the two atomic bombs before the August 1945 bombings of Hiroshima and Nagasaki. Warner was the chief of the assembly group that put together the Fat Man plutonium bomb.

Warner, who was now Division Leader of Z-Division of the Manhattan District (Operation Crossroads), wanted a bomb to be used as a dress rehearsal prior to one of the Bikini Atoll atomic tests. So in 1947 Les Rowe saw an opportunity to dispose of Bomb #3. With its nuclear core removed, the bomb was flown over the ocean and dropped about 500 miles west of Los Angeles.

Bikini Atoll

In his book *Operation Crossroads: The Atomic Tests at Bikini Atoll*, Jonathan M. Weisgall quotes Bob Hope as saying, "as soon as the war ended, we located the one spot on earth that hadn't been touched by the war and blew it to hell." Prior to the tests at Bikini Atoll, the U.S. Navy moved the atoll's 167 residents to the Rongerik Atoll. This atoll was one-sixth the size of Bikini Atoll. Furthermore, the Navy left them an inadequate supply of water and food. However, the Bikinians wanted to start a new life and planted new crops and began fishing. They soon found that the land produced fewer crops and the sea produced fewer fish.

In early 1948, the islanders were on the verge of starvation. This time they were moved to Kwajalein, about 200 miles southeast of Bikini.

The U.S. government allowed 160 Bikini islanders to return to Bikini in 1970. In 1978, blood tests showed an 11-fold increase in caesium-137 as well as higher levels of plutonium-239 and strontium-90.

The Bikinians won a class-action suit against the United States and received $75 million in damages and the creation of a trust fund of $90 million.

Wendover Air Base, Utah

Although much of Wendover Air Base is still a restricted area, the nonprofit public foundation Historic Wendover Airfield (HWA) has restored many of the important historical sites.

The airfield is the most original remaining and operating World War II training airfield in the United States. No other airfield has surpassed the historic importance of those at Wendover. The HWA has also renovated multiple hangers (including the one that housed the *Enola Gay*) and restored many of the barracks and support buildings. Visitors to these sites will gain an appreciation of the dedicated work and sacrifice by the 509th Composite Group.

In 1984, I received permission to visit the restricted part of Wendover Air Base. I enjoyed walking along the airstrips and seeing the original assembly areas where "Little Boy" and "Fat Man" were housed for delivery to Tinian. Personally,

Chapter 14. Epilogue: Japan

Photograph of Samuel S. Kloda with one of the "Little Boy" practice bombs, taken in 1984 at Wendover Air Force Base (author's collection).

On August 25, 1990, members and friends of the 509th Composite Group erected a beautiful 16-foot memorial at Wendover, Utah. Autographed by Gilbert B. Dickman, the pilot who delivered "Fat Man" from Wendover to Tinian Island (author's collection).

finding some of the "practice bombs" that were used by the 509th during their practice runs was the best part of my visit.

On August 25, 1990, members and friends of the 509th Composite Group erected a beautiful 16-foot memorial monument at Wendover, Utah. The inscription on the 16-foot peace includes these words:

Chapter 14. Epilogue: Japan

This monument is dedicated to members of the 509th Composite Group, United States Army Air Force, who trained at the Wendover, Utah Army Air Force Base in 1944–45 for the vital, secret mission of delivering the first atomic bombs on Japanese targets in August 1945.... Recognition is given to the scientific teams who created this awesome weapon, those who sacrificed their lives in the Pacific Theater, Allied forces in other theaters of action during World War II, and all who contributed to bring this dreadful war to an end.

The loss of lives of the people of Hiroshima and Nagasaki, Japan are especially recognized in this memorial for their sacrifice to mankind's struggle for a more peaceful world.

MEMORIAL INSCRIPTION

In Memory

This monument is dedicated to the members of the 509th Composite Group, United States Army Air Force, who trained at the Wendover, Utah Army Air Force Base in 1944-45 for the vital, secret mission of delivering the first atomic bombs on Japanese targets in August 1945. The combined efforts of all members of the United States Army and the United States Navy who created the massive armed service support of this historic endeavor share this important dedication that brought World War II to a much earlier conclusion.

Recognition is given to the scientific teams who created this awesome weapon, those who sacrificed their lives in the Pacific Theatre, Allied forces in other theatres of action during World War II, and all who contributed to bring this dreadful war to an end.

The loss of lives of the people of Hiroshima and Nagasaki, Japan are especially recognized in this memorial for their sacrifice to mankind's struggle for a more peaceful world.

May this monument stand as a symbol of hope that mankind will reason and work together for the ultimate goal of world peace.

Erected by Members and Friends of the
509th Composite Group
August 25, 1990

Memorial inscription. Autographed by James Van Pelt (author's collection).

Group photo at memorial dedication. *Left to right:* **Jacob Beser, Theodore Van Kirk, Thomas Ferebee, Richard Nelson, Paul Tibbets, George Caron. Autographed by Theodore Van Kirk (author's collection).**

> May this monument stand as a symbol of hope that mankind will reason and work together for the ultimate goal of world peace.

Truman

About one year after the end of World War II, there were many individuals who started to criticize the United States, and especially President Truman, for dropping two atomic bombs. They believed that Japan was almost defeated and on the verge of collapse. After the fall of Okinawa, all military outposts were under U.S. control. The United States fire bombed cities without any real opposition.

On October 14, 1948, in Milwaukee, Wisconsin, President Truman delivered a nationwide radio broadcast:

> As President of the United States, I had the fateful responsibility of deciding whether or not to use this weapon for the first time. It was the hardest decision I ever had to make. But the President cannot duck hard problems—he cannot pass the buck.

Opposite: **509th Memorial dedication. Copy of program given to invited guests, including Japanese visitors from Hiroshima and Nagasaki. Autographed by Fred Bock (author's collection).**

Chapter 14. Epilogue: Japan

I made the decision after discussions with the ablest men in our Government, and after long and prayerful consideration.

I decided that the bomb should be used in order to end the war quickly and save countless lives—Japanese as well as American. What this meant for the future was staggering to think about.

In his memoirs (1955), Truman continued to defend the use of atomic weapons to end the war. He told a journalist that "it was done to save 125,000 youngsters on the U.S. side and 125,000 on the Japanese side from getting killed and that is what it did. It probably also saved a half million youngsters on both sides from being maimed for life."

President Harry S. Truman died on December 26, 1972, at the age of 88.

Soviet Union

During World War II, all the major play-

509th pamphlet. The U.S. Post Office issued a special cancellation. Autographed by Leonard Cheshire (author's collection).

Party at B-29 *Enola Gay* hangar, Wendover, Utah (1990). *Left to right:* **Samuel S. Kloda, Sandy O'Hara ("Miss Pin-Up, Vietnam Conflict"), Paul Tibbets. Autographed by Sandy O'Hara and Paul Tibbets (author's collection).**

ers—Britain, Germany, the Soviet Union and the United States—had spies operating everywhere. Germany and the Soviet Union were especially interested in the scientists working in Los Alamos.

To counteract this espionage, the United States employed thousands of military officers and the FBI. However, the Soviets were able, through

President Harry S. Truman signs the proclamation designating August 1, 1946, as Air Force Day. With him are General Carl A. Spaatz, commanding general of the Army Air Forces, and Lieutenant General Ira C. Eaker, deputy commander of the AAF. Autographed by General Eaker (U.S. Air Force; author's collection).

different individuals, to collect enough data to build their own atomic bomb.

On August 29, 1949, the Soviet Union conducted its first nuclear test (code named "RDS-1") at the Semipalatinsk test site in modern-day Kazakhstan. The device had a yield of 22 kilotons.

It was no coincidence that the RDS-1 device bore a close resemblance to the "Fat Man" bomb dropped on Nagasaki, as Soviet espionage had managed to obtain details about the U.S. Manhattan Project and the Trinity test on July 16, 1945. The Soviet device was therefore a plutonium-based implosion device.

It is estimated that 130,000 men and women worked on the Manhattan Project. Soviet espionage directed at the Manhattan Project started in September 1941, almost one year before the Manhattan Engineer District was created. The KGB's code name for the American project was ENORMOZ ("enormous").

Soviet intelligence agencies in Moscow pressured their American "friends" to develop sources within the Manhattan Project. Most of their recruits were Americans who were sympathetic to communism or members of the American Communist Party. In many cases, these individuals who chose to provide

JOHN EDGAR HOOVER
DIRECTOR

Federal Bureau of Investigation
United States Department of Justice
Washington, D. C.

February 9, 1945

SECRET

PERSONAL AND CONFIDENTIAL
BY SPECIAL MESSENGER

Honorable Harry L. Hopkins
The White House
Washington, D. C.

Dear Harry:

       As you are well aware, the Army for the past two years has been vitally interested in a highly secret project for the development of an atomic explosive. This explosive, as you know, involves the release of energy through the shattering of atoms of heavy elements.

       During the period that the Army has been engaged in the supervision of this experimentation, numerous efforts have been made by the Soviets to obtain the highly secret information concerning the experimentation and this Bureau has been actively following such Soviet efforts.

       The German Government has also been interested in the same type of experimentation in Germany and has attempted to get information regarding the atomic experimentation in the United States.

       Recently, in connection with the operation of a radio station by a German agent under control of the Federal Bureau of Investigation but which station the Germans believe to be a free station, an inquiry was received from Germany containing the following questions regarding the status of atomic explosive experimentation in the United States:

       First, where is heavy water being produced? In what quantities? What method? Who are users?

       Second, in what Laboratories is work being carried on with large quantities of uranium? Did accidents happen there? What does the protection against Neutronic Rays consist of in these Laboratories? What is the material and the strength of coating?

DECLASSIFIED AND RELEASED
E.O. 11652, Sec. 3(E) and 5(D) or (E)
Authority FBI Ltr. May 29, 1975
By _____ NARS, Date _____
(NND 750152)

Above and next page: On February 9, 1945, FBI Director J. Edgar Hoover wrote a letter to Secretary of Commerce Harry L. Hopkins. Hoover was worried about Soviet and German interest in atomic explosive experiments (Franklin D. Roosevelt Presidential Library and Museum).

information to the Soviet Union did so for purely ideological reasons, not for money.

In January 1950, the British physicist Klaus Fuchs was discovered to have been a Soviet spy. Fuchs had worked at Los Alamos in the Theoretical Division and passed information regarding atomic weapons design to the Soviets.

The most famous "atomic spies" were Julius and Ethel Rosenberg. Ethel

> Third, is anything known concerning the production of bodies or molecules out of metallic uranium rods, tubes, plates? Are these bodies provided with coverings for protection? Of what do these coverings consist?
>
> We have already advised the appropriate authorities in the War Department concerning these German inquiries.
>
> I thought the foregoing would be of considerable interest to the President.
>
> With best wishes and kind regards,
>
> Sincerely yours,
>
> Edgar

Rosenberg's brother was David Greenglass. He was an army machinist who worked on the Manhattan Project. He was briefly stationed at the Clinton Engineer Works uranium enrichment facility at Oak Ridge, Tennessee, and then worked at the Los Alamos Laboratory in New Mexico from August 1944 until February 1946. He was a spy for the Soviet Union and funneled vital information regarding the atomic bomb to Julius Rosenberg.

In 1950, David Greenglass was caught and, because he cooperated with the government, given a 15-year jail sentence. Julian and Ethel Rosenberg refused to cooperate or even admit any participation in aiding the Soviets. They were executed on June 19, 1953.

In the mid–1990s, another scientist was discovered to have passed secrets to the Soviet Union. Theodore Hall was a physicist at Los Alamos and supplied information on the implosion design.

Unbelievable as this may seem, a fourth major spy who had worked at Los Alamos was discovered after 70 years. Harvey Klehr and John Earl Haynes wrote a brilliant and well-researched essay in the September 2019 issue of *Studies in Intelligence: On the Trail of a Fourth Soviet Spy at Los Alamos* (the CIA's in-house journal concerning Oscar Seborer).

In October 1944, Seborer was assigned to the Oak Ridge complex in Tennessee as an electrical engineer. During the Trinity atomic test in Alamogordo, New Mexico, Seborer was "part of a unit monitoring seismological effects" of the first detonation of the atomic device. The essay quotes a member of Seborer's contacts as saying, "He handed over to them the formula for the 'A' bomb."

The Arms Race

After the Bikini Atoll tests in 1946, the United States and the Soviet Union rushed to develop more powerful bombs.

The first hydrogen bomb was developed by a team of American scientists led by Edward Teller and Stanislaw Ulam. On November 1, 1952, this new nuclear weapon was tested. The test was part of "Operation Ivy," and the nuclear weapon (code named "Mike") had a yield of 10.4 megatons. (One megaton is roughly equivalent to 1 million tons of TNT.) The Soviet Union followed this American test by testing its first hydrogen bomb on August 12, 1953.

The first U.S. air drop of a thermonuclear weapon occurred on May 20, 1956. The pilot, Major David M. Critchlow, dropped a hydrogen bomb (TX-15-X1) from a B-52B-30-BO Stratofortress at an altitude of 50,000 feet. The explosive force of the TX-15 was rated at 3.8 megatons. The observers confirmed that the fireball caused by the explosion measured at least four miles in diameter and was brighter than the light from 500 suns. The mushroom cloud rose to 94,000 feet.

The test was intended to collect weapon effects data for high-yield air bursts. However, it also made a political demonstration of the United States' ability to deliver H-bombs by air to anywhere in the world.

This test and others that followed caused concerns among scientists and environmentalists about the effects of radioactive fallout on human and animal life. The United States began to push for a ban on open-air atomic testing. In addition, the Disarmament Committee of the United Nations (United States, the United Kingdom, Canada, France, and the Soviet Union) started to negotiate an international agreement to end nuclear testing.

On August 5, 1963, the Test Ban Treaty was signed in Moscow and ratified by the U.S. Senate on September 24, 1963. The treaty prohibited nuclear tests in the atmosphere, in outer space, and under water.

Nuclear weapons test "Mike" (yield was 10.4 megatons) on Eniwetok Atoll. The test was part of "Operation Ivy." "Mike" was the first hydrogen bomb (H-bomb) to ever be tested (National Nuclear Security Administration/Nevada Site Office).

Hirohito

After Japan's surrender, Emperor Hirohito was allowed to stay on the

On May 21, 1956, Major David M. Critchlow, flying a Boeing RB-52B-30-BO Stratofortress assigned to the 4925th Test Group (Atomic), Kirtland Air Force Base, Albuquerque, New Mexico, took off from Eniwetok Island ("Fred Island"), the main island of Eniwetok Atoll in the Marshall Islands, carrying a TX-15-X1 two-stage radiation implosion thermonuclear bomb weighing 6,867 pounds. The bomb was approximately 136 inches long, with a diameter of 34.5 inches. The target was a point on Namu Island, Bikini Atoll. It was dropped from 50,000 feet and exploded 4,350 feet over the open ocean. The explosive force of the TX-15 was rated at 3.8 megatons. Autographed by David M. Critchlow (U.S. Air Force; author's collection).

University of California, Berkeley, Lawrence Berkeley National Laboratory. *Left to right:* Edwin McMillan, director, Lawrence Lab; Glenn Seaborg, chairman, Atomic Energy Commission; President John F. Kennedy; Edward Teller; Robert McNamara, secretary of defense. Autographed by Glenn Seaborg, Edward Teller, and Robert McNamara (John F. Kennedy Library; author's collection).

throne. General Douglas MacArthur had insisted that Hirohito retain his position as emperor. MacArthur saw the emperor as a symbol of the continuity and cohesion of the Japanese people. The postwar Japanese constitution preserved the monarchy but defined Hirohito as a figurehead with no political power. All political power went to elected representatives.

From 1945 to 1951, Hirohito toured the country and oversaw reconstruction efforts. After the U.S. occupation, Hirohito served largely in the background while Japan went through a period of rapid economic growth.

Hirohito was not indicted as a war criminal, in part because U.S. authorities feared it could throw their occupation into chaos. Hirohito would later portray himself as a powerless constitutional monarch. However, the Official Imperial Constitution adopted under Emperor Meiji gave full power to the emperor. Article 4 stated, "The Emperor is the head of the Empire, combining in Himself the rights of sovereignty, and exercises them, according to the provisions of the present Constitution," while, according to article 6, "The Emperor gives sanction to laws and orders them to be promulgated and executed." Article 11 likewise states, "The Emperor has the supreme command of the Army and the Navy." The emperor was thus the leader of the Imperial General Headquarters.

Emperor Hirohito played an active role in military decisions. Besides the unprovoked attack on Pearl Harbor, Hirohito was directly responsible for many other atrocities committed by his military. Here we cite only a few:

1. The "Rape of Nanking" was an episode of mass murder and rape committed by Japanese troops against the residents of Nanjing (Nanking), then the capital of China, during the Second Sino-Japanese War. The massacre occurred over a period of six weeks starting on December 13, 1937. Japanese army officers initiated the so-called Manchurian Incident by detonating a railway explosion and blaming it on Chinese bandits. They then used the event as an excuse to take over Manchuria in northeastern China and set up a puppet state there.

2. Soldiers of the Imperial Japanese Army murdered at least 200,000 Chinese civilians and perpetrated widespread rape and looting. (China's official estimate is more than 300,000 died, based on the evaluation of the Nanjing War Crimes Tribunal in 1947.)

3. In the Battle of Wuhan (from August to October 1938), poison gas weapons, such as phosgene, were produced by Unit 731 and authorized by specific orders given by Hirohito himself, transmitted by the chief of staff of the army. The emperor authorized the use of toxic gas on 375 separate occasions, despite the resolution adopted by the League of Nations on May 14 condemning Japanese use of toxic gas.

4. After the April 9, 1942, U.S. surrender of the Bataan Peninsula on the main Philippine island of Luzon to the Japanese, approximately 75,000 Filipino and American prisoners of war were forced to make a strenuous 65-mile march to prison camps. During the march, prisoners received little food or water, and many died. They were subjected to severe physical abuse, including beatings and torture. On the march, the "sun treatment" was a common form of torture: Prisoners were forced to sit in sweltering direct sunlight without helmets or other head coverings. Anyone who asked for water was shot dead. Some men were told to strip naked or sit within sight of fresh, cool water. Trucks drove over some of those who fell or succumbed to fatigue, and "clean-up crews" put to death those too weak to continue. Some marchers were randomly stabbed with bayonets or beaten.

Although this "Bataan Death March" occurred in April 1942, the U.S. government waited until January 27, 1944, to inform the American public about the march. It then released sworn statements of military officers who had escaped.

The Bataan Death March and other Japanese actions were used to arouse fury in the United States. General George Marshall made the following statement:

> These brutal reprisals upon helpless victims evidence the shallow advance from savagery which the Japanese people have made.... We serve notice upon the Japanese military and political leaders as well as the Japanese people that the future of the Japanese race itself,

depends entirely and irrevocably upon their capacity to progress beyond their aboriginal barbaric instincts.

Emperor Hirohito died on January 7, 1989, having spent nearly 64 years on the throne. He was the longest-living emperor to reign over Japan.

Paul Tibbets

After leaving Roswell, New Mexico, Tibbets attended the Air Command and Staff School at Maxwell Air Force Base, Alabama. In 1947, he was posted to the Directorate of Requirements at Air Force Headquarters at the Pentagon. Soon after, Tibbets was appointed director of the Directorate of Requirements Strategic Air Division, which was responsible for drawing up requirements for future bombers. Based on his extensive experience, Tibbets believed that the bombers of the future would be jet aircraft. He became involved in the Boeing B-47 Stratojet program.

The Boeing B-47 Stratojet was a long-range, six-engine, turbojet-powered strategic bomber designed to fly at high subsonic speed and at high altitude to avoid enemy interceptor aircraft. The primary mission of the B-47 was serving as a nuclear bomber capable of striking targets within the Soviet Union.

Tibbets became the B-47 project officer at Boeing in Wichita from July 1950

Colonel Paul Tibbets as test pilot of the B-47 bomber in 1951. Autographed by Paul Tibbets (author's collection).

Paul W. Tibbets, the pilot of the *Enola Gay* (the airplane that dropped the atomic bomb on Hiroshima), speaking at the Harry S. Truman Library. He (center) addressed the crowd at the Harry S. Truman Appreciation Society program. Later he attended a book signing in the museum. Representatives of Whiteman Air Force Base were in the background, as were members of the American Legion Band, who performed earlier. Donor: Ed Ireland (Harry S. Truman Library & Museum).

Enola Gay with a new paint job. Autographed by Paul Tibbets, Theodore Van Kirk and Thomas Ferebee (U.S. Air Force; author's collection).

until February 1952. He then became commander of the Proof Test Division at Eglin Air Force Base in Valparaiso, Florida, where flight testing of the B-47 was conducted.

In 1954, Tibbets returned to the Air War College in Montgomery, Alabama. In June 1955, he became director of war plans at the Allied Air Forces in Central

Europe at Fontainebleau, France. He returned to the United States in February 1956 to command the 308th Bomb Wing at Savannah, Georgia.

In January 1958, Tibbets became commander of the 6th Air Division at Mac-Dill Air Force Base, Florida, and was promoted to brigadier general in 1959. This assignment was followed by another tour of duty at the Pentagon as director of management analysis. In July 1962, he was assigned to the Joint Chiefs of Staff as deputy director for operations and then, in June 1963, as deputy director for

Enola Gay, **Boeing B-29 Superfortress, in the National Air and Space Museum's Steven F. Udvar-Hazy Center in Virginia, part of the Smithsonian (Mys 721tx, Wikimedia Commons).**

Bockscar, **Boeing B-29 Superfortress, at the United States Air Force Museum in Dayton, Ohio (USAF Museum, Wikimedia Commons).**

the National Military Command System. After leaving Washington, Tibbets was named military attaché in India. He retired from the United States Air Force on August 31, 1966. After leaving the Air Force, he worked for Executive Jet Aviation, serving on the founding board and as its president from 1976 until his retirement in 1987.

Throughout his lifetime, Tibbets defended the use of atomic bombs and gave speeches to discourage their future use. Tibbets believed that he had saved millions of lives. He never had remorse for his place in history. He credited the use of two atomic bombs with ending the war and saving millions of lives, both American and Japanese.

When asked about his participation in the atomic strikes, Tibbets would always say, "I am an airman, a pilot. In 1945, I was wearing the uniform of the U.S. following the orders of our commander-in-chief."

This is the official photograph of Brigadier General Paul Tibbets. Autographed by Paul Tibbets (U.S. Air Force; author's collection).

Today, the *Enola Gay* is on permanent display at the National Air and Space Museum's Steven F. Udvar-Hazy Center in Virginia. *Bockscar* is on permanent display at the United States Air Force Museum in Dayton, Ohio.

Brigadier General Paul W. Tibbets, Jr., died on November 1, 2007, at the age of 92. He had requested that his ashes be scattered over the English Channel, where he did many hours of flying during World War II. His grandson, Lieutenant Colonel Paul Tibbets IV, a B-2 Command pilot, fulfilled his request.

Bibliography

Akizuki, Tatsuichiro. *Nagasaki 1945: The First Full-Length Eyewitness Account of the Atomic Bomb Attack on Nagasaki*. London: Quartet Books, 1981.

Alperovitz, Gar. *Atomic Diplomacy—Hiroshima and Potsdam: The Use of the Atomic Bomb and the American Confrontation with Soviet Power*. New York: Simon & Schuster, 1965.

Andrade, Edward N. da C. *Rutherford and the Nature of the Atom*. London: Heinemann Educational Books, 1964.

Arnold, Henry H. *Global Mission*. 1949. Reprint, New York: Arno Press, 1972.

Baker, Paul R. *The Atomic Bomb: The Great Decision*: New York: Holt, Rinehart and Winston, 1968.

Beser, Jacob. *Hiroshima & Nagasaki*. Memphis, TN: Global Press, 1988.

Bess, Michael. *Choices Under Fire: Moral Dimensions of World War II*. New York: Vintage Books, 2006.

Bush, Vannevar. *Modern Arms and Free Men*. New York: Simon & Schuster, 1949.

Chang, Iris. *The Rape of Nanking: The Forgotten Holocaust of World War II*. New York: Penguin, 1997.

Chappell, David. *Before the Bomb: How America Approached the End of the Pacific War*. Lexington: University Press of Kentucky, 1996.

Chinnock, Frank. *Nagasaki: The Forgotten Bomb*. New York: World, 1969.

Churchill, Winston S. *Triumph and Tragedy*. Boston: Houghton Mifflin, 1953.

Cimino, Al. *The Manhattan Project: The Making of the Atomic Bomb*. London: Arcturus Publishing, 2015.

Department of State. *Foreign Relations of the United States: Diplomatic Papers, the Potsdam Conference, 1945*. Vols. I and II. Washington, DC: U.S. Government Printing Office, 1960.

Groueff, Stephane. *Manhattan Project: The Untold Story of the Making of the Atomic Bomb*. Boston: Little, Brown, 1967.

Groves, Leslie R. *Now It Can Be Told: The Story of the Manhattan Project*. New York: Harper & Brothers, 1962.

Hachiya, Michihiko. *Hiroshima Diary: The Journal of a Japanese Physician, August 6–September 30, 1945—Fifty Years Later*. Chapel Hill: University of North Carolina Press, 1995.

Hamby, Alonzo L. *Man of the People: A Life of Harry S. Truman*. Oxford: Oxford University Press, 1995.

Hawkins, David. *Manhattan District History: Project Y, the Los Alamos Project*. Vol. I. Los Alamos, NM: University of California and U.S. Atomic Energy Commission, 1945.

Hecht, Selig. *Explaining the Atom*. New York: Alfred A. Knopf, 1947.

Hersey, John R. *Hiroshima*. New York: Alfred A. Knopf, 1946.

Hubert, Richard S. R. "The OWI Saipan Operation." Official Report to U.S. Information Service, Washington, DC, 1946.

James, David H. *The Rise and Fall of the Japanese Empire*. New York: Macmillan, 1951.

Jungk, Robert. *Brighter Than a Thousand Suns*. New York: Harcourt, Brace, 1958.

Klehr, Harvey, and John Earl Haynes. "On the Trail of a Fourth Soviet Spy at Los Alamos." *Studies in Intelligence* 63, no. 3 (September 2019).

Lamont, Lansing. *Day of Trinity*. New York: Atheneum, 1965.

Laurence, William L. *Dawn Over Zero: The Story of the Atomic Bomb*. New York: Alfred A. Knopf, 1946.

LeMay, Curtis E., with M. Kantor. *Mission with LeMay*. New York: Doubleday, 1965.

Los Alamos National Laboratory. *Los Alamos: The Beginning of an Era, 1943–1945*. Los Alamos, NM: Self-published, 1967.

MacArthur, Douglas. *Reminiscences*. New York: McGraw-Hall, 1964.

Morton, Louis. "The Decision to Use the Atomic Bomb." In *Command Decisions*. Office of the Chief of Military History, Department of the Army. Washington, DC: U.S. Government Printing Office, 1960.

Nagai, Takashi. *We of Nagasaki*. New York: Duell, Sloan and Pearce, 1951.

Nelson, Craig. *The Age of Radiance: The Epic Rise and Dramatic Fall of the Atomic Era*. New York: Scribner, 2014.

Olivi, Fred J., and William R. Watson Jr. *Decision at Nagasaki: The Mission That Almost Failed*. Self-published, 1998.

O'Reilly, Bill. *The Day the World Went Nuclear*. New York: Henry Holt, 2017.

Raymond, Robert. *Out of the Fiery Furnace: The Impact of Metals on the History of Mankind*. University Park: Pennsylvania State University Press, 1990.

Reilly, Michael F. *I Was Roosevelt's Shadow*. London: W. Foulsham, 1946.

Rhodes, Richard. *The Making of the Atomic Bombs*. New York: Simon & Schuster, 1986.

Robertson, John Kellock. *Atomic Artillery and the Atomic Bomb*. New York: D. Van Nostrand, 1945.

Rotter, Andrew J. *Hiroshima: The World's Bomb*. New York: Oxford University Press, 2008.

Rowe, James Les. *Project W-47*. Livermore, CA: Jā A Rō Publishing, 1972.

Russ, Harlow W. *Project Alberta: The Preparation of Atomic Bombs for Use in World War II*. Los Alamos, NM: Exceptional Books, 1984.

Smaltz, Donald C. "Independent Counsel: A View from Inside." *Georgetown Law Journal* (July 1998).

Smyth, H. D. *A General Account of the Development of Methods of Using Atomic Energy for Military Purposes Under the Auspices of the United States Government, 1940–1945*. Princeton, NJ: Princeton University Press, 1948.

Southard, Susan. "Opinion: Nagasaki, the Forgotten City." *New York Times*, August 7, 2015.

Stimson, Henry L. *On Active Service in Peace and War*. New York: Harper & Brothers, 1948.

Teller, Edward, and Allen Brown. *The Legacy of Hiroshima*. 1962. Reprint, Westport, CT: Greenwood Press, 1975.

Terkel, Louis. "One Hell of a Big Bang." *The Guardian*, August 6, 2002.

Thomas, Gordon, and Max Morgan Witts. *Enola Gay*. New York: Stein and Day, 1977.

Tibbets, Paul W., with Clair Stebbins and Harry Franken. *The Tibbets Story*. New York: Stein and Day, 1978.

Truman, Harry S. *The Autobiography of Harry S. Truman*. Edited by Robert H. Ferrell. Columbia: University of Missouri Press, 2002.

———. *Memoirs: Year of Decisions*. Garden City, NY: Doubleday, 1955.

Trumbull, Robert. *Nine Who Survived Hiroshima and Nagasaki*. New York: E. P. Dutton, 1957.

U.S. Joint Task Force One. *Bombs at Bikini*. New York: W. H. Wise, 1947.

Waxman, Olivia B. "How the U.S. and Japan Became Allies Even After Hiroshima and Nagasaki." *Time*, August 6, 2018.

Weisgall, Jonathan M. *Operation Crossroads: The Atomic Tests at Bikini Atoll*. Arlington, VA: Naval Institute Press, 1994.

White Sands Missile Range, Public Affairs Office. *Trinity Site: 1945–1995*. New Mexico: Good Press, 2019.

William, Josette H. *The Information War in the Pacific, 1945*. Washington, DC: Central Intelligence Agency, 2009.

Index

Numbers in ***bold italics*** indicate pages with illustrations

"Able" 162, 164–165, ***171***
Adams, William B. 163
Aeby, Jack 40–41, ***160***
Agnew, Harold ***19***, 33, 34, 85, 91, 127
Aioi Bridge 100, 106, ***181***
Akizuki, Tatsuichiro 138
Albury, Charles D. ***51***, 53, ***82, 124***, 127–128, 133–134, 138–139, 142, 144, 145
Allison, Samuel K. ***19***, 27, 31–32
Alvarez, Luis 22, 27, ***29***, 85, 91, 127
Ashworth, Frederick L. 27, ***29***, 122, 129, 135
Atomic Bomb Dome (Genbaku Dome) 110, 182–183

Bacher, Robert F. ***23***, 25, 27, 35–36
Badali, Joseph ***95***
Bainbridge, Kenneth T. 27, 30–32
"Baker" 162, 167, 169–170, ***172–178***
Barnes, Philip M. 129
"Bataan Death March" 202
Battle of Wuhan, 202
Beahan, Kermit 53, 128, 133, 135, ***140, 143, 147***
Beser, Jacob 53, 54, 66, 88, 90, 93, ***97***, 98, 128–132, ***145, 150, 183, 192***
Bethe, Hans ***22***, 25, 38
Big Stink (later called *Dave's Dream*) 126, 130, 138, 167
Bikini Atoll 122, 163, 166–167, 171, 174, 188
Birch, Francis 27
Blandy, William, P. 162, 164
Bock, Frederick ***81***, 126, 135–136, ***145, 151***, 193
Bockscar 81, 82, 126, 129–131, 134, 135, 137–138, 143, 145, ***206***
Bohr Niels ***6***, 7, 22, ***61***
Bradbury, Norris 22, 27, ***31***, 35
Briggs, Lyman J. 10
Brixner, Berlyn 33–37, ***40***
Brown, Harold ***23***

Buckley, Edward 128
Buscher, Joseph ***88***
Bush, Vannevar 30
Byrnes, James F. ***17***

Caron, George 52, 80, ***84***, 93, 95, ***97***, 101, ***102–103***, 110, ***121, 192***
Chadwick, James 7, 30
"Charlie" 162, 173
Cheshire, Leonard 128, ***194***
Choices Under Fire: Moral Dimensions of World War II (Bess) 105
Chugoka Shimbun building 111, 113
Churchill, Winston 41–42, ***43***, 68
Clark, Mark 46, ***49***
Classen, Thomas 52, ***121***
Cockroft, John 7
code names: Alamogordo, New Mexico test ("Trinity") 27; Allied Invasion of Northwest Africa ("Operation Torch") 46; Atomic Bomb ("Gadget") 24, 30, 31; Authority to Requisition ("Silverplate") 27, 51; Battle of Okinawa ("Operation Iceberg") 59; Clinton Pile Graphine Reactor ("X-10") 26; Fire Bombing of Tokyo ("Operation Meetinghouse") 56; Hiroshima Bomb ("Little Boy") 26; Honshu Island Attack ("Coronet") 58; Implosion Experimental Group ("E-5") ***31***; Japanese Occupation) ("Operation Blacklist") 179; Japan's Unconditional Surrender ("Operation Downfall") 58, 67; KGB-American Project ("ENORMOZ") 196; Kyushu Island Attack ("Olympic") 58; Large Inert Bomb Copies ("Pumpkins") 58; Los Alamos Laboratory ("Project Y") 21; Nagasaki Bomb ("Fat Man") 26; Operation Ivy ("MIKE") 199–200; Optics ("Group

G-11") 38; Ordnance Division ("E-7 Group") 22; Original Gun-Type Plutonium Weapon ("Thin-Man") 26; Project Alberta ("Project A") 27, 127–128; Special Bomb Operational Order No. 13 ("Centerboard") 77; Tibbets ("Dimples 82") 77; 216th Army Air Force Base Unit (Project W-47") 27, 187

Dave's Dream 164, 166, 168–170
Davies, James 114–115, ***121, 153***
Decision at Nagasaki: The Mission That Almost Failed (Olivi) 137
Dehart, Albert T. 128, ***133***
Dickman, Gilbert B. 126, ***190***
Dike, Sheldon 22, 27
Doll, Edward E. ***29***, 88, 129
Doolittle, James H. 47–48
Downey, William B. ***84–87***
Duzenbury, Wyatt ***9***, 52, ***84***, 90, ***97***

Eaker, Ira C. ***47, 196***
Eatherly, Claude 82, 98
Einstein, Albert 7, ***8, 9, 61***
Eisenhower, Dwight D. 48
Enola Gay (No. 82) 78, ***80–81***, 82, ***83***, 85, 88–89, 93–98, 110, 119–120, 127, 129, 188, ***206***, 207
Ent, Uzal G. 50
Executive Jet Aviation Inc. 79

Faillace, Gaetano 180
Farrell, Thomas F. ***29***, 32, 38, 69–70, 110, 114, 129, 143
"Fat Man" 126, ***127–128***, 129–130, 163, 190, 196
Ferebee, Tom ***9***, ***52–54, 56***, 66, 77, ***81, 84***, 90, 93, 96, ***97–98, 114, 119***, 132, 135, ***164, 192, 205***
Fermi, Enrico 7, ***10***, 11, 17, ***19***, 21, 22, 31, 33, 91
Feynmann, Richard 22
Foster, John 23
Franck, James 59–60, ***61***

Index

Franck Report 59, **60**
Frisch, Otto 22, **24**
Fuchs, Klaus 197
Full House 82, 89

Gallagher, Ray 128–129, 133, **147**
Giles, Barnet M. 121
Goodman, Walter 127
Great Artiste 82, 85, 88, 96, 98, 126–127, 130–131, 135
Greenglass, David 198
Greisen, Kenneth 30, 32
Groves, Leslie R. 16, **19**, 22, 25, 30, **39**, 41, 50, 55–56, 58, 66, 94, 101, 135, 185

Hahn, Otto 5, 7
Hall, Theodore 199
Halsey, William F. 115
Handy, Thos. T. 66
Harrison, William C. 163, 165, **168**, 176–178
Hashimoto, Mochitsura 69
Hirohito, Emperor 123, 154–158, 179, **180**, 200–203
Hiroshima Diary (Hachiya) 106
Hiroshima: The World's Bomb (Rotter) 105
Holloway, Marshall 30
Hoover, J. Edgar 197–198
Hope, Bob 188
Hopkins, Harry L. 197
Hopkins, James **121**, 126, 130, 138
Hubbard, J.M. 30

USS *Indianapolis* 65, 69–70, 71, 72
Instrument of Surrender **157–159**
Iwo Jima 82, 85, 96

Jabbitt III 82, 89
Jeppson, Morris **9**, 56, 79–80, 93, 96, **183**
Johnson, Lawrence 29, 85, 91, 127

Kelsou, Nakao 72
Kennedy, John F. **23, 201**
Kennedy, Joseph w. 25
King, John A. **62–63**
Kistiakowsky, George 22, 27, 30–32
Klaproth, Martin Heinrich 5
Kokura 127, 130–131, 133
Kuharek, John D. 128–129, **133**
Kupferberg, Jess 127–128
Kyushu 138

Laggin' Dragon 127
Landshoff, Rolf 22
Lansdale, John 50
Lansing State Journal 70
Laurence, William "Bill" 30, 38, 85, 127
Lawrence, Ernest O. **21, 25**, 26, 30

Leahy, W.D. 117
LeMay, Curtis 65, 77, 115, 126, 129, 143, 162, 187
"LeMay Bombing Leaflets **72–75**
Lemnitzer, Lyman 46–47, **48**
Levy, Charles **88**
Lewis, Robert **52**, 56, **84**, 90, **97**, **104–105, 119, 121**, 184
"Little Boy" 78, 82–83, 85, 94, **95**, 101, 106, 182
Los Alamos 21–23

MacArthur, Douglas 67, 158, **159**, 179, **180**, 201
Mack, Julian 38
The Making of the Atomic Bomb (Rhodes) 64
Manhattan Project: Advisory Committee 10, 11; creation 16; costs 17; project members 22
Marquardt, George 82, 85, 91, 96, 98, **107**, 127, 130, **145**
Marshall, George 66, 185, 202
McKibben, Joe 30–31
McKnight, Chuck 82, 96, 127
McMillan, Edwin 22
McNamara, Robert 23, **201**
McVay, Charles Butler III 65
Meitner, Lisa 7
USS *Missouri* 157–159, 161
Mitsubishi Steel and Arms Works 143–144
Mitsubishi-Urakami Ordance Works 137
Morrison, Philip 27, 64
Muranaka, Keiichi 75

Nagasaki Genbakuki (Nagasaki 1945) (Akizuki) 138
Nelson, Richard **84**, 93, 95, **97**, **115**, 127, 133, 135–146, **192**
USS *Nevada* 165, 167
Nimitz, Chester 67, 115
No. 91 82, 85, 88, 96, 98, **107**
Nolan, James F. 27
Norstad, Lauris 48, **49**, 148
Now It Can Be Told (Groves) 50

O'Hara, Sandy **195**
Okamura, Michinobu 73
Olivi, Fred **81, 124**, 128, 133–134, **137**
Operation Crossroads 162–164, 166–167, **169–172**, 178, 188
Operation Crossroads: The Atomic Test at Bikini Atoll (Weisgall) 188
Operation Ivy 199–200
Oppenheimer, J. Robert **16**, 22, **25**, 27, **28**, 32, **39**, 54, 101
Out of the Fiery Furnace (Raymond) 38

Parsons, William S. "Deak" 22, 25, 27, **29**, 50, 65, 79, 82, 85, 87–88, 93–94, 96, **97**, 110, 115, **121**, 126, 129

Payette, Hazen 53, 115, **121**
Pearl Harbor 12–13, 123
Penny, William 27
Percival, Arthur E. 159
plutonium 16, 26, 170
plutonium bomb 30, 126
Porter, John **121**
Potsdam Conference 41
Potsdam Proclamation 68–69, 154
Price, James N. **64**
Project Albert 27, 64, 85, 91, 129, 145
proximity fuze 94, 101
Purnell, William R. **29, 54**, 65, 114, **121**, 126

Rabi, Isidor I. **21**, 22, 32, **61**
Ramsey, Norman F. 22, 27, **29**, 50, 85, 87
Rosenberg, Ethel 197, 199
Rosenberg, Julius 197–199
Roosevelt, Franklin D. **8, 9**, 12, **17**; "Day of Infamy Speech" **13–15**, 16, 28, 56; "Rape of Nanking" 202
Rossi, Bruno 22
Rowe, James Lee 186–187
Rutherford, Ernest **5**, 7

Sachs, Alexander 8, 9, **10**, 11
Seaborg, Glenn T. 16, **19**, 23, 25, 59–60, 170, **201**
Seborer, Oscar 199
Segre, Emilo 22, 26
Semple, David 167
Serber, Robert 27, **29**
Shigemitsu, Mamoru 157–158
Shumard, Robert H. **84**, 93, **97**
"Silverplate" 27, 51
Spaatz, Carl A. **48**, 66–67, 110, 114–115, **121**, 126, 143, **196**
Special Bombing Mission No. 13 77
Spitzer, Abe 128, 133, **136**
Stalin, Joseph 41, **43**, 68
Stiborik, Joe **84**, 93, **97, 108, 181**
Stimson, Henry 12, 56–58, 118, 155
Straight Flush 82, 89, 98
Strassmann, Fritz 7
Studies In Intelligence (Klehr and Haynes) 199
Swancutt, Woodrow P. 163–164, **167, 169–170, 172**
Sweeney, Charles "Chuck" 53, 82, 88, 96, 98, **108**, 126, 128–129, 133–137, 141, **145**, 147, **149, 153**, 184
Sziland, Leo 7, **8, 9–10**, 11, **12, 19**, 31, 59–60

Taylor, Ralph 82
Teller, Edward 22, **23**, 32, 41, 199, **201**
Thelen, Dick 70, **71**
Thurlow, Setsuko 184
Tibbets, Paul, Jr. **9, 52–54**, 62,

66, *84*, *93*, *97*, *120–122*, *129*, *163–164*, *195*, *203–205*, *207*; Bikini Atoll tests 162–163; Bizette attack 47–49; bombing of Hiroshima 94, 95–101; commander, 509th Composite Group 50; commander of the 340th Bomb Squadron 46; *Enola Gay* crew 90–93; key group 53; Nagasaki bombing 129, 135–138; Operation Torch 46; planning Hiroshima attack 77–78, 82, 84–85, 88–89; returning to Tinian 119–122
Tibbets, Paul IV 207
Tinian 62–75
Tokyo 56
Top Secret 82, 96
Trinitite (Alamogordo glass) 38
Trinity Project *30–42*
Trinity Site: 1945–1995 38

Truman, Bess 156
Truman, Harry S. *155–156*, *196*; Japan surrenders 154, 155–159; Manhattan Project 57–58; 1948 radio broadcast 192, 194; at Potsdam 41–45; Potsdam Declaration 68–69; swearing-in as president 56; Truman's address to the nation 123, 125
Tully, Grace 12
Twining, Nathan 114–115, *121*

Uanna, William "Bud" 53
Ulam, Stanislaw 199
Umezu, Yoshijiro 158–159
uranium bomb 25–26
Urey, Harold C. *19*

Van Kirk, Theodore "Dutch" *9*, *52–54*, 56, *66*, *81*, *84*, 90, 92–93, *97*, *109*, *119*, *121*, *124*, *192*, *205*
Van Pelt, James 53, 92, 128, 133–134, *145*, *147*, *152*, 191
von Neumann, John 22

Wainwright, Johnathan 159
Waldman, Bernard 22, 27, *29*, 85, 91
Walton, Ernest 7
Warner, Roger 27, *29*, 186, 188
Wendover (Utah) Air Base 51, 53–55, 129, 184, 186–188
Wigner, Eugene 31
Wilson, John 82
Wood, Harold H. 163

Yakushima 127, 130
Young, Don 89

Zinn, Walter 7, *19*